Differential Equations: A Beginner's Guide

DIFFERENTIAL EQUATIONS:
A BEGINNER'S GUIDE

Arun Dubey

Oxford Book Company
Jaipur ▪ New Delhi

Published By

Oxford Book Company

Regd. Off. "Bony Residency", Gate No. 2
Opp. Tilak Public School, Vishweswariya Nagar,
Gopalpura Road, Jaipur - 302018, India
Ph: 0141 - 2761280, Telefax: 2761381
email - oxfordbookjpr@gmail.com
website - oxfordbookcompany.com

Delhi Office: 102, 1st Floor, Satyam House,
4327/3, Ansari Road, Darya Ganj,
New Delhi - 110002, India
Ph: 011-45652440

Distributed By
KEYA INTERNATIONAL PRESS LLP
E-mail: keyainternationalpressllp@gmail.com

First published in India, 2026

ISBN : 978-93-5524-489-5

Typeset by Shivangi Computers, Jaipur

Printed and bound in India by Thomson Press Ltd.,
New Delhi

Contents

Preface

Differential equations form the cornerstone of mathematical modeling, bridging the gap between abstract theory and real-world applications. From predicting population growth to understanding fluid dynamics and engineering complex systems, differential equations serve as an indispensable tool across disciplines. However, for many beginners, the subject often appears daunting, filled with intricate formulas and abstract concepts. This book, Differential Equations: A Beginner's Guide, is designed to break down those barriers and present the subject in a clear, accessible, and engaging manner.

This book is crafted with the needs of students, educators, and self-learners in mind, offering a structured introduction to differential equations without overwhelming the reader with unnecessary complexity. Starting with fundamental concepts, it gradually builds up to more advanced techniques, ensuring a smooth and logical progression. Each chapter is supplemented with intuitive explanations, step-by-step solutions, and real-world examples that illustrate how differential equations shape various scientific and engineering domains.

One of the key objectives of this book is to demystify differential equations by focusing on their practical significance. Whether it is modeling natural phenomena such as heat transfer, oscillations, or financial markets, understanding how differential equations operate in different contexts provides a deeper appreciation of their power and versatility. Through a combination of theoretical discussions and applied problem-solving, this guide ensures that readers develop both conceptual clarity and computational proficiency.

To enhance the learning experience, numerous solved examples, exercises, and graphical interpretations are included, catering to different learning styles. The book also introduces numerical methods, helping readers appreciate the role of computational tools in solving complex differential equations that may not have analytical solutions. Additionally, a concise historical perspective is provided to highlight the evolution of differential equations and their contributions to scientific advancements.

Differential Equations: A Beginner's Guide is a thorough and interesting introduction to differential equations, regardless of whether you are a student just starting out in the subject, an instructor searching for a well-organized teaching resource, or a professional looking for a refresher. After finishing this book, readers will have a solid understanding of the underlying ideas and the self-assurance to use differential equations in their particular professions.

— ***Arun Dubey***

Chapter 1

Fractional Calculus

Fractional calculus is a branch of mathematical analysis that studies the possibility of taking real number powers or complex number powers of the differentiation operator.

$$D = \frac{d}{dx},$$

and the integration operator J. (Usually J is used instead of I to avoid confusion with other I-like glyphs and identities.)

In this context the term *powers* refers to iterative application or composition, in the same sense that $f^2(x) = f(f(x))$. For example, one may ask the question of meaningfully interpreting

$$\sqrt{D} = D^{\frac{1}{2}}$$

as a square root of the differentiation operator (an operator half iterate), i.e., an expression for some operator that when applied *twice* to a function will have the same effect as differentiation. More generally, one can look at the question of defining

$$D^a$$

for real-number values of a in such a way that when a takes an integer value n, the usual power of n-fold differentiation is recovered for $n > 0$, and the $-n$th power of J when $n < 0$.

The motivation behind this extension to the differential operator is that the semigroup of powers D^a will form a *continuous* semigroup with parameter a, inside which the original *discrete* semigroup of D^n for integer n can be recovered as a subgroup. Continuous semigroups are prevalent in mathematics, and have an interesting theory. Notice here that *fraction* is then a misnomer for the exponent a, since it need not be rational; the use of the term *fractional calculus* is merely conventional. Fractional differential equations are a generalisation of differential equations through the application of fractional calculus.

Nature of the Fractional Derivative

An important point is that the fractional derivative at a point x is a *local property* only when a is an integer; in non-integer cases we cannot say that the fractional derivative at x of a function f depends only on the graph of f very near x, in the way that integer-power derivatives certainly do. Therefore it is expected that the theory involves some sort of boundary conditions, involving information on the function further out. To use a metaphor, the fractional derivative requires some peripheral vision.

As far as the existence of such a theory is concerned, the foundations of the subject were laid by Liouville in a paper from 1832. The fractional derivative of a function to order a is often now defined by means of the Fourier or Mellin integral transforms.

Riemann–Liouville Fractional Integral

The classical form of fractional calculus is given by the Riemann–Liouville integral, essentially what has been described above. The theory for periodic functions, therefore including the 'boundary condition' of repeating after a period, is the Weyl integral. It is defined on Fourier series, and requires the constant Fourier coefficient to vanish (so, applies to functions on the unit circle integrating to 0).

$$_aD_t^{-\alpha} f(t) = aI_t^{\alpha} f(t) = \frac{1}{\Gamma(\alpha)} \int_a^t (t-\tau)^{\alpha-1} f(\tau)d\tau$$

By contrast the Grünwald–Letnikov derivative starts with the derivative instead of the integral.

Riemann–Liouville Fractional Derivative

The corresponding derivative is calculated using Lagrange's rule for differential operators. Computing n-th order derivative over the integral of order n-α, the α order derivative is obtained. Is important to remark that n is the nearest integer bigger than α.

$$ {}_aD_t^\alpha f(t) = \frac{d^n}{dt^n}\, aD_t^{-(n-\alpha)} f(t) = \frac{d^n}{dt^n}\, aI_t^{n-\alpha} f(t) $$

Caputo Fractional Derivative

There is another option for computing fractional derivatives; the Caputo fractional derivative. It was introduced by M. Caputo in 1990. In contrast to the Riemann Liouville fractional derivative, when solving differential equations using Caputo's definition, it is not necessary to define the fractional order initial conditions. Caputo's definition is illustrated as follows.

$$ {}_a^cD_t^\alpha f(t) = \frac{1}{\Gamma(n-\alpha)} \int_a^t \frac{f^{(n)}(\tau)d\tau}{(t-\tau)^{\alpha+1-n}} $$

Functional Calculus

In the context of functional analysis, functions *f(D)* more general than powers are studied in the functional calculus of spectral theory. The theory of pseudo-differential operators also allows one to consider powers of *D*. The operators arising are examples of singular integral operators; and the generalisation of the classical theory to higher dimensions is called the theory of Riesz potentials. So there are a number of contemporary theories available, within which *fractional calculus* can be discussed.

Applications

Fractional Conservation of Mass: As described by Wheatcraft and Meerschaert (2008), a fractional conservation of mass equation is needed when the control volume is not large enough compared to the scale of heterogeneity and when the flux within the control volume is non-linear. In the referenced paper, the fractional conservation of mass equation for fluid flow is:

$$ -\rho(\nabla^\alpha \cdot \vec{u}) = \Gamma(\alpha+1)\Delta x^{1-\alpha}\rho(\beta_s + \phi\beta_w)\frac{\partial p}{\partial t} $$

Fractional Advection Dispersion Equation: This equation has been shown useful for modelling contaminant flow in heterogenous porous media.

Structural Damping Models: Fractional derivatives are used to model viscoelastic damping in certain types of materials like polymers.

Linear Differential Equation

In mathematics, linear differential equations are differential equations that have solutions which can be added together to form other solutions. They can be ordinary or partial. The solutions to linear equations form a vector space (not the case with non-linear differential equations).

Linear differential equations are of the form

$$Ly = f$$

where the differential operator L is a linear operator, y is the unknown function (such as a function of time $y(t)$), and the right hand side f is a given function of the same nature as y (called the source term). For a function dependent on time we may write the equation more expressively as

$$Ly(t) = f(t)$$

and, even more precisely by bracketing

$$L[y(t)] = f(t)$$

The linear operator L may be considered to be of the form

$$L_n(y) \equiv \frac{d^n y}{dt^n} + A_1(t)\frac{d^{n-1}y}{dt^{n-1}} + \cdots + A_{n-1}(t)\frac{dy}{dt} + A_n(t)y$$

The linearity condition on L rules out operations such as taking the square of the derivative of y; but permits, for example, taking the second derivative of y. It is convenient to rewrite this equation in an operator form

$$L_n(y) \equiv \left[D^n + A_1(t)D^{n-1} + \cdots + A_{n-1}(t)D + A_n(t) \right] y$$

where D is the differential operator d/dt (i.e. $Dy = y'$, $D^2y = y$¬,...), and the A_n are given functions.

Such an equation is said to have order n, the index of the highest derivative of y that is involved.

A typical simple example is the linear differential equation used to model radioactive decay. Let N(t) denote the number of radioactive atoms in some sample of material at time t. Then for some constant k > 0, the number of radioactive atoms which decay can be modelled by

$$\frac{dN}{dt} = -kN$$

If y is assumed to be a function of only one variable, one speaks about an ordinary differential equation, else the derivatives and their coefficients must be understood as (contracted) vectors, matrices or tensors of higher rank, and we have a (linear) partial differential equation.

The case where $f = 0$ is called a homogeneous equation and its solutions are called complementary functions. It is particularly important to the solution of the general case, since any complementary function can be added to a solution of the inhomogeneous equation to give another solution (by a method traditionally called *particular integral and complementary function*). When the A_i are numbers, the equation is said to have *constant coefficients*.

Homogeneous Equations with Constant Coefficients

The first method of solving linear ordinary differential equations with constant coefficients is due to Euler, who realised that solutions have the form e^{zx}, for possibly-complex values of z.

The exponential function is one of the few functions that keep its shape after differentiation. In order for the sum of multiple derivatives of a function to sum up to zero, the derivatives must cancel each other out and the only way for them to do so is for the derivatives to have the same form as the initial function. Thus, to solve

$$y^{(n)} + A_1 y^{(n-1)} + \cdots + A_n y = 0$$

we set $y = e^{zx}$, leading to

$$z^n e^{zx} + A_1 z^{n-1} e^{zx} + \cdots + A_n e^{zx} = 0.$$

Division by e^{zx} gives the nth-order polynomial

$$F(z) = z^n + A_1 z^{n-1} + \cdots + A_n = 0.$$

This algebraic equation $F(z) = 0$, is the characteristic equation considered later by Gaspard Monge and Augustin-Louis Cauchy.

Formally, the terms

$$y^{(k)} \qquad (k=1,2,\ldots,n).$$

of the original differential equation are replaced by z^k. Solving the polynomial gives n values of z, z_1, ..., z_n. Substitution of any of those values for z into e^{zx} gives a solution $e^{z_i x}$. Since homogeneous linear differential equations obey the superposition principle, any linear combination of these functions also satisfies the differential equation.

When these roots are all distinct, we have n distinct solutions to the differential equation. It can be shown that these are linearly independent, by applying the Vandermonde determinant, and together they form a basis of the space of all solutions of the differential equation.

The preceding gave a solution for the case when all zeros are distinct, that is, each has multiplicity 1. For the general case, if z is a (possibly complex) zero (or root) of $F(z)$ having multiplicity m, then, for $k \in \{0,1,\ldots,m-1\}$, $y = x^k e^{zx}$ is a solution of the ODE.

Applying this to all roots gives a collection of n distinct and linearly independent functions, where n is the degree of $F(z)$. As before, these functions make up a basis of the solution space.

If the coefficients A_i of the differential equation are real, then real-valued solutions are generally preferable. Since non-real roots z then come in conjugate pairs, so do their corresponding basis functions $x^k e^{zx}$, and the desired result is obtained by replacing each pair with their real-valued linear combinations Re(y) and Im(y), where y is one of the pair.

A case that involves complex roots can be solved with the aid of Euler's formula.

Nonhomogeneous Equation with Constant Coefficients

To obtain the solution to the nonhomogeneous equation (sometimes called inhomogeneous equation), find a particular integral $y_p(x)$ by either the method of undetermined coefficients or the method of variation of parameters; the general solution to the linear differential equation is the sum of the general solution of the related homogeneous equation and the particular integral.

Or, when the initial conditions are set, use Laplace transform to obtain the particular solution directly. Suppose we face

$$\frac{d^n y(x)}{dx^n} + A_1 \frac{d^{n-1} y(x)}{dx^{n-1}} + \cdots + A_n y(x) = f(x).$$

For later convenience, define the characteristic polynomial

$$P(v) = v^n + A_1 v^{n-1} + \cdots + A_n.$$

We find the solution basis $\{y_1(x), y_2(x), \ldots, y_n(x)\}$ as in the homogeneous (*f(x)=0*) case. We now seek a particular integral $y_p(x)$ by the variation of parameters method. Let the coefficients of the linear combination be functions of x:

$$y_p(x) = u_1(x)y_1(x) + u_2(x)y_2(x) + \cdots + u_n(x)y_n(x).$$

For ease of notation we will drop the dependency on x (i.e. the various *(x)*). Using the operator notation $D = d/dx$, the ODE in question is $P(D)y = f$; so

$$f = P(D)y_p = P(D)(u_1 y_1) + P(D)(u_2 y_2) + \cdots + P(D)(u_n y_n).$$

With the constraints

$$0 = u_1' y_1 + u_2' y_2 + \cdots + u_n' y_n$$

$$0 = u_1' y_1' + u_2' y_2' + \cdots + u_n' y_n'$$

$$\cdots$$

$$0 = u_1' y_1^{(n-2)} + u_2' y_2^{(n-2)} + \cdots + u_n' y_n^{(n-2)}$$

the parameters commute out,

$$f = u_1 P(D) y_1 + u_2 P(D) y_2 + \cdots + u_n P(D) y_n + u_1' y_1^{(n-1)} + u_2' y_2^{(n-1)} + \cdots + u_n' y_n^{(n-1)}.$$

But $P(D)y_j = 0$, therefore

$$f = u_1' y_1^{(n-1)} + u_2' y_2^{(n-1)} + \cdots + u_n' y_n^{(n-1)}.$$

This, with the constraints, gives a linear system in the $u2_j$. This much can always be solved; in fact, combining Cramer's rule with the Wronskian,

$$u_j' = (-1)^{n+j} \frac{W(y_1, \ldots, y_{j-1}, y_{j+1} \ldots, y_n)_{\binom{0}{f}}}{W(y_1, y_2, \ldots, y_n)}.$$

The rest is a matter of integrating $u2_j$.

The particular integral is not unique; $y_p + c_1 y_1 + \cdots + c_n y_n$ also satisfies the ODE for any set of constants c_j.

Equation with Variable Coefficients

A linear ODE of order n with variable coefficients has the general form

$$p_n(x)y^{(n)}(x) + p_{n-1}(x)y^{(n-1)}(x) + \cdots + p_0(x)y(x) = r(x).$$

Examples

A simple example is the Cauchy–Euler equation often used in engineering

Nonlinear Partial Differential Equations

In mathematics and physics, nonlinear partial differential equations are (as their name suggests) partial differential equations with nonlinear terms. They describe many different physical systems, ranging from gravitation to fluid dynamics, and have been used in mathematics to solve problems such as the Poincaré conjecture and the Calabi conjecture. They are difficult to study: there are almost no general techniques that work for all such equations, and usually each individual equation has to be studied as a separate problem.

Methods for Studying Nonlinear Partial Differential Equations

Existence and Uniqueness of Solutions: A fundamental question for any PDE is the existence and uniqueness of a solution for given boundary conditions. For nonlinear equations these questions are in general very hard: for example, the hardest part of Yau's solution of the Calabi conjecture was the proof of existence for a Monge–Ampere equation.

Singularities: The basic questions about singularities (their formation, propagation, and removal, and regularity of solutions) are the same as for linear PDE, but as usual much harder to study. In the linear case one can just use spaces of distributions, but nonlinear PDEs are not usually defined on arbitrary distributions, so one replaces spaces of distributions by refinements such as Sobolev spaces.

An example of singularity formation is given by the Ricci flow: Hamilton showed that while short time solutions exist, singularities will usually form after a finite time. Perelman's solution of the Poincaré conjecture depended on a deep study of these singularities, where he showed how to continue the solution past the singularities.

Linear Approximation: The solutions in a neighbourhood of a known solution can sometimes be studied by linearising the PDE around the solution. This corresponds to studying the tangent space of a point of the moduli space of all solutions.

Moduli Space of Solutions: Ideally one would like to describe the (moduli) space of all solutions explicitly, and for some very special PDEs this is possible. (In general this is a hopeless problem: it is unlikely that there is any useful description of all solutions of the Navier–Stokes equation for example, as this would involve describing all possible fluid motions.)

If the equation has a very large symmetry group, then one is usually only interested in the moduli space of solutions modulo the symmetry group, and this is sometimes a finite dimensional compact manifold, possibly with singularities; for example, this happens in the case of the Seiberg–Witten equations.

A slightly more complicated case is the self dual Yang–Mills equations, when the moduli space is finite dimensional but not necessarily compact, though it can often be compactified explicitly. Another case when one can sometimes hope to describe all solutions is the case of completely integrable models, when solutions are sometimes a sort of superposition of solitons; for example, this happens for the Korteweg–de Vries equation.

Exact Solutions: It is often possible to write down some special solutions explicitly in terms of elementary functions (though it is rarely possible to describe all solutions like this).

One way of finding such explicit solutions is to reduce the equations to equations of lower dimension, preferably ordinary differential equations, which can often be solved exactly. This can sometimes be done using separation of variables, or by looking for highly symmetric solutions.

Some equations have several different exact solutions.

Numerical Solutions: Numerical solution on a computer is almost the only method that can be used for getting information about arbitrary systems of PDEs. There has been a lot of work done, but a lot of work still remains on solving certain systems numerically, especially for the Navier–Stokes and other equations related to weather prediction.

Lax Pair: If a system of PDEs can be put into Lax pair form

$$\frac{dL}{dt} = LA - AL$$

then it usually has an infinite number of first integrals, which help to study it.

Euler–Lagrange Equations: Systems of PDEs often arise as the Euler–Lagrange equations for a variational problem. Systems of this form can sometimes be solved by finding an extremum of the original variational problem.

Hamilton Equations

Integrable Systems

PDEs that arise from integrable systems are often the easiest to study, and can sometimes be completely solved. A well known example is the Korteweg–de Vries equation.

Symmetry

Some systems of PDEs have large symmetry groups. For example, the Yang–Mills equations are invariant under an infinite dimensional gauge group, and many systems of equations (such as the Einstein field equations) are invariant under diffeomorphisms of the underlying manifold. Any such symmetry groups can usually be used to help study the equations; in particular if one solution is known one can trivially generate more by acting with the symmetry group. Sometimes equations are parabolic or hyperbolic "modulo the action of some group": for example, the Ricci flow equation is not quite parabolic, but is "parabolic modulo the action of the diffeomorphism group", which implies that it has most of the good properties of parabolic equations.

Attributes of Variables

Dependent and Independent Variables

The terms "dependent variable" and "independent variable" are used in similar but subtly different ways in mathematics and statistics as part of the standard terminology in those subjects. They are used to distinguish between two types of quantities being considered, separating them into those available at the start of a process and those being created by it, where the latter (dependent variables) are dependent on the former (independent variables).

Simplified Example

The independent variable is typically the variable representing the value being manipulated or changed and the dependent variable is the observed result of the independent variable being manipulated. For example concerning nutrition, the independent variable of daily vitamin C intake (how much vitamin C one consumes) can influence the dependent variable of life expectancy (the average age one attains). Over some period of time, scientists will control the vitamin C intake in a substantial group of people.

One part of the group will be given a daily high dose of vitamin C, and the remainder will be given a placebo pill (so that they are unaware of not belonging to the first group) without vitamin C.

The scientists will investigate if there is any statistically significant difference in the life span of the people who took the high dose and those who took the placebo (no dose). The goal is to see if the independent variable of high vitamin C dosage has a correlation with the dependent variable of people's life span.

Use in Mathematics

In calculus, a function is a map whose action is specified on variables. Take x and y to be two variables. A function f may map x to some expression in x. Assigning $y = f(x)$ gives a relation between y and x. If there is some relation specifying y in terms of x, then y is known as a "dependent variable" (and x is an "independent variable").

Use in Statistics

Controlled Experiments: In a statistics experiment, the dependent variable is the event studied and expected to change whenever the independent variable is altered.

In the design of experiments, an independent variable's values are controlled or selected by the experimenter to determine its relationship to an observed phenomenon (i.e., the dependent variable). In such an experiment, an attempt is made to find evidence that the values of the independent variable determine the values of the dependent variable. The independent variable can be changed as required, and its values do not represent a problem requiring explanation in an analysis, but are taken simply as given. The dependent variable, on the other hand, usually cannot be directly controlled.

Controlled variables are also important to identify in experiments. They are the variables that are kept constant to prevent their influence on the effect of the independent variable on the dependent. Every experiment has a controlling variable, and it is necessary to not change it, or the results of the experiment won't be valid.

"Extraneous variables" are those that might affect the relationship between the independent and dependent variables. Extraneous variables are usually not theoretically interesting. They are measured in order for the experimenter to compensate for them. For example, an experimenter who wishes to measure the degree to which caffeine intake (the independent variable) influences explicit recall for a word list (the dependent variable) might also measure the participant's age (extraneous variable). She can then use these age data to control for the uninteresting effect of age, clarifying the relationship between caffeine and memory.

In summary:

- Independent variables answer the question "What do I change?"
- Dependent variables answer the question "What do I observe?"
- Controlled variables answer the question "What do I keep the same?"

- Extraneous variables answer the question "What uninteresting variables might mediate the effect of the IV on the DV?"

Alternative Terminology in Statistics: In statistics, the dependent/independent variable terminology is used more widely than just in relation to controlled experiments. For example the data analysis of two jointly varying quantities may involve treating each in turn as the dependent variable and the other as the independent variable. However, for general usage, the pair response variable and explanatory variable is preferable as quantities treated as "independent variables" are rarely statistically independent.

Depending on the context, an independent variable is also known as a "predictor variable," "regressor," "controlled variable," "manipulated variable," "explanatory variable," "exposure variable," and/or "input variable." A dependent variable is also known as a "response variable," "regressand," "measured variable," "observed variable," "responding variable," "explained variable," "outcome variable," "experimental variable," and/or "output variable."

In addition, some special types of statistical analysis use terminology more relevant to the specific context. For example reliability theory uses the term exposure variable for what would otherwise be an explanatory or independent variable; medical statistics may use the term risk factor; and machine learning and pattern recognition use the term feature.

Examples:

- If one were to measure the influence of different quantities of fertiliser on plant growth, the independent variable would be the amount of fertiliser used (the changing factor of the experiment). The dependent variables would be the growth in height and/or mass of the plant (the factors that are influenced in the experiment) and the controlled variables would be the type of plant, the type of fertiliser, the amount of sunlight the plant gets, the size of the pots, etc. (the factors that would otherwise influence the dependent variable if they were not controlled).

- In a study of how different doses of a drug affect the severity of symptoms, a researcher could compare the frequency and intensity of symptoms (the dependent variables) when different doses (the independent variable) are administered, and attempt to draw a conclusion.
- In measuring the acceleration of a vehicle, time is usually the independent variable, while speed is the dependent variable. This is because when taking measurements, times are usually predetermined, and the resulting speed of the vehicle is recorded at those times.

  As far as the experiment is concerned, the speed is dependent on the time. Since the decision is made to measure the speed at certain times, time is the independent variable.
- In measuring the amount of colour removed from beetroot samples at different temperatures, the dependent variable would be the amount of pigment removed, since it is depending on the temperature (which is the independent variable).
- In sociology, in measuring the effect of education on income or wealth, the dependent variable could be a level of income or wealth measured in monetary units (United States Dollars for example), and an independent variable could be the education level of the individual(s) who compose(s) the household (i.e. academic degrees).

Manipulated versus Subject Variables

Up to this point, the term independent variable has meant some factors that are manipulated directly by the researcher. An experiment compares one condition created by and under the control of the experimenter with another.

In many studies, however, comparisons are also made between groups of people who differ from each other in ways other than those manufactured by the person designing the study.

These comparisons are made between factors that are referred to variously as:

- Extraneous variables (Subject variables, Experimental variables, Situational variables)
- Pseudo-independent variables
- Ex post facto variables
- Natural group variables
- Non-manipulated variables

They refer to already existing characters of the subject participating in the study, such as gender, age, intelligence, physical or psychiatric disorder, and any personality attribute and any personality attribute one can name. When using subject variables in a study, the researcher cannot manipulate them directly but must *select* subjects for the different conditions by virtue of the characteristics they already have.

Chapter 2

Fundamental Theorem of Calculus

The *fundamental theorem of calculus* is the statement that differentiation and integration are inverse operations: if a continuous function is first integrated and then differentiated, the original function is retrieved. An important consequence, sometimes called the *second fundamental theorem of calculus*, allows one to compute integrals by using an antiderivative of the function to be integrated.

Statements of Theorems

- *Fundamental theorem of calculus.* Let f be a continuous real-valued function defined on a closed interval $[a, b]$. Let F be the function defined, for all x in $[a, b]$, by

$$F(x) = \int_a^x f(t)dt.$$

Then, F is continuous on $[a, b]$, differentiable on the open interval (a, b), and

$$F'(x) = f(x)$$

for all x in (a, b).

- *Second fundamental theorem of calculus.* Let f be a real-valued function defined on a closed interval $[a, b]$ that admits an antiderivative g on $[a, b]$. That is, f and g are functions such that for all x in $[a, b]$,

$$f(x) = g'(x).$$

If f is integrable on $[a, b]$ then

$$\int_a^b f(x)dx = g(b) - g(a).$$

Extensions

Improper Integrals

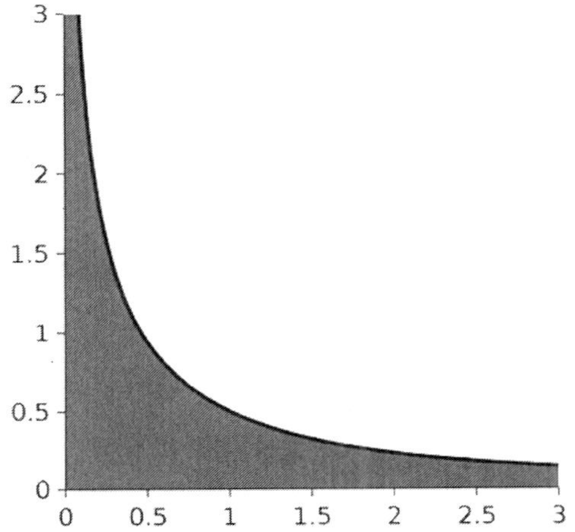

Figure: *The improper integral*

$$\int_0^{\infty} \frac{dx}{(x+1)\sqrt{x}} = \pi$$

has unbounded intervals for both domain and range.

A "proper" Riemann integral assumes the integrand is defined and finite on a closed and bounded interval, bracketed by the limits of integration. An improper integral occurs when one or more of these conditions is not satisfied. In some cases such integrals may be defined by considering the limit of a sequence of proper Riemann integrals on progressively larger intervals.

If the interval is unbounded, for instance at its upper end, then the improper integral is the limit as that endpoint goes to infinity.

$$\int_a^{\infty} f(x)dx = \lim_{b\to\infty} \int_a^b f(x)dx$$

If the integrand is only defined or finite on a half-open interval, for instance $(a,b]$, then again a limit may provide a finite result.

$$\int_a^b f(x)dx = \lim_{\epsilon\to 0} \int_{a+\epsilon}^b f(x)dx$$

That is, the improper integral is the limit of proper integrals as one endpoint of the interval of integration approaches either a specified real number, or ∞, or $-\infty$. In more complicated cases, limits are required at both endpoints, or at interior points.

Consider, for example, the function $1/((x+1)\sqrt{x})$ integrated from 0 to " (shown right). At the lower bound, as x goes to 0 the function goes to ∞, and the upper bound is itself ∞, though the function goes to 0.

Thus this is a doubly improper integral. Integrated, say, from 1 to 3, an ordinary Riemann sum suffices to produce a result of $\pi/6$. To integrate from 1 to ∞, a Riemann sum is not possible.

However, any finite upper bound, say t (with $t > 1$), gives a well-defined result, $2\arctan(\sqrt{t})-\pi/2$. This has a finite limit as t goes to infinity, namely $\pi/2$. Similarly, the integral from 1/3 to 1 allows a Riemann sum as well, coincidentally again producing $\pi/6$. Replacing 1/3 by an arbitrary positive value s (with $s < 1$) is equally safe, giving $\pi/2-2\arctan(\sqrt{s})$.

This, too, has a finite limit as s goes to zero, namely $\pi/2$. Combining the limits of the two fragments, the result of this improper integral is

$$\begin{aligned}
\int_0^{\infty}\frac{dx}{(x+1)\sqrt{x}} &= \lim_{s\to 0}\int_s^1\frac{dx}{(x+1)\sqrt{x}}+\lim_{t\to\infty}\int_1^t\frac{dx}{(x+1)\sqrt{x}} \\
&= \lim_{s\to 0}\left(\frac{\pi}{2}-2\arctan\sqrt{s}\right)+\lim_{t\to\infty}\left(2\arctan\sqrt{t}-\frac{\pi}{2}\right) \\
&= \frac{\pi}{2}+\left(\pi-\frac{\pi}{2}\right) \\
&= \frac{\pi}{2}+\frac{\pi}{2} \\
&= \pi.
\end{aligned}$$

This process does not guarantee success; a limit may fail to exist, or may be unbounded. For example, over the bounded interval 0 to 1 the integral of $1/x$ does not converge; and over the unbounded interval 1 to ∞ the integral of $1/\sqrt{x}$ does not converge.

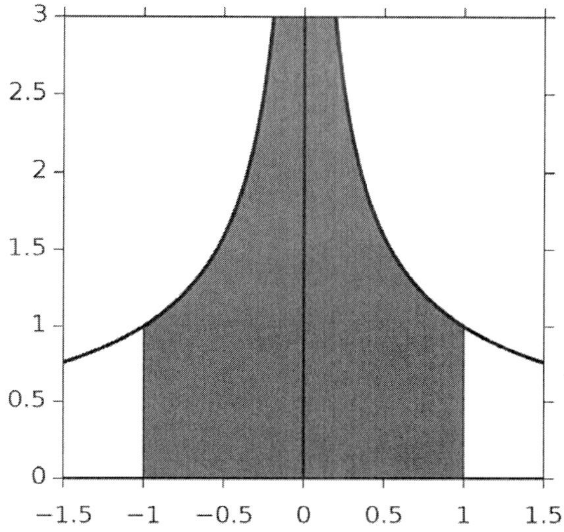

Figure: *The improper integral*

$$\int_{-1}^{1} \frac{dx}{\sqrt[3]{x^2}} = 6$$

is unbounded internally, but both left and right limits exist.

It may also happen that an integrand is unbounded at an interior point, in which case the integral must be split at that point, and the limit integrals on both sides must exist and must be bounded. Thus

$$\begin{aligned}\int_{-1}^{1} \frac{dx}{\sqrt[3]{x^2}} &= \lim_{s\to 0}\int_{-1}^{-s} \frac{dx}{\sqrt[3]{x^2}} + \lim_{t\to 0}\int_{t}^{1} \frac{dx}{\sqrt[3]{x^2}} \\ &= \lim_{s\to 0} 3(1-\sqrt[3]{s}) + \lim_{t\to 0} 3(1-\sqrt[3]{t}) \\ &= 3+3 \\ &= 6.\end{aligned}$$

But the similar integral

$$\int_{-1}^{1} \frac{dx}{x}$$

cannot be assigned a value in this way, as the integrals above and below zero do not independently converge.

Multiple Integration

Integrals can be taken over regions other than intervals. In general, an integral over a set E of a function f is written:

$$\int_E f(x)dx.$$

Here x need not be a real number, but can be another suitable quantity, for instance, a vector in R^3. Fubini's theorem shows that such integrals can be rewritten as an *iterated integral*. In other words, the integral can be calculated by integrating one coordinate at a time.

Just as the definite integral of a positive function of one variable represents the area of the region between the graph of the function and the x-axis, the *double integral* of a positive function of two variables represents the volume of the region between the surface defined by the function and the plane which contains its domain. (The same volume can be obtained via the *triple integral* — the integral of a function in three variables — of the constant function $f(x, y, z) = 1$ over the above mentioned region between the surface and the plane.) If the number of variables is higher, then the integral represents a hypervolume, a volume of a solid of more than three dimensions that cannot be graphed.

For example, the volume of the cuboid of sides 4 × 6 × 5 may be obtained in two ways:

By the double integral $\iint_D 5\,dxdy$ of the function $f(x, y) =$ 5 calculated in the region D in the xy-plane which is the base of the cuboid. For example, if a rectangular base of such a cuboid is given via the xy inequalities 3 d– x d– 7, 4 d– y d– 10, our above double integral now reads

$$\int_4^{10}\left[\int_3^7 5\,dx\right]dy$$

From here, integration is conducted with respect to either x or y first; in this example, integration is first done with respect to x as the interval corresponding to x is the inner integral. Once the first integration is completed via the $F(b) - F(a)$ method or otherwise, the result is again integrated with respect to the other variable. The result will equate to the volume under the surface.

By the triple integral $\iiint_{\text{cuboid}} 1dxdydz$ of the constant function 1 calculated on the cuboid itself.

Line Integrals

The concept of an integral can be extended to more general domains of integration, such as curved lines and surfaces. Such integrals are known as line integrals and surface integrals respectively. These have important applications in physics, as when dealing with vector fields.

A *line integral* (sometimes called a *path integral*) is an integral where the function to be integrated is evaluated along a curve. Various different line integrals are in use. In the case of a closed curve it is also called a *contour integral.*

The function to be integrated may be a scalar field or a vector field. The value of the line integral is the sum of values of the field at all points on the curve, weighted by some scalar function on the curve (commonly arc length or, for a vector field, the scalar product of the vector field with a differential vector in the curve). This weighting distinguishes the line integral from simpler integrals defined on intervals. Many simple formulas in physics have natural continuous analogs in terms of line integrals; for example, the fact that work is equal to force, F, multiplied by displacement, s, may be expressed (in terms of vector quantities) as:

$$W = \vec{F} \cdot \vec{s}.$$

For an object moving along a path in a vector field $\vec{F}$ such as an electric field or gravitational field, the total work done by the field on the object is obtained by summing up the differential work done in moving from $\vec{s}$ to $\vec{s} + d\vec{s}$. This gives the line integral

$$W = \int_C \vec{F} \cdot d\vec{s}.$$

Surface Integrals

A *surface integral* is a definite integral taken over a surface (which may be a curved set in space); it can be thought of as the double integral analog of the line integral. The function to be integrated may be a scalar field or a vector field. The value of the surface integral is the sum of the field at all points on the surface. This can be achieved by splitting the surface into surface elements, which provide the partitioning for Riemann sums.

For an example of applications of surface integrals, consider a vector field v on a surface S; that is, for each point x in S, $v(x)$ is a vector. Imagine that we have a fluid flowing through S, such that v(x) determines the velocity of the fluid at x. The flux is defined as the quantity of fluid flowing through S in unit amount of time. To find the flux, we need to take the dot product of v with the unit surface normal to S at each point, which will give us a scalar field, which we integrate over the surface:

$$\int_S \mathbf{v} \cdot d\mathbf{S}.$$

The fluid flux in this example may be from a physical fluid such as water or air, or from electrical or magnetic flux. Thus surface integrals have applications in physics, particularly with the classical theory of electromagnetism.

Integrals of Differential Forms

A differential form is a mathematical concept in the fields of multivariable calculus, differential topology and tensors. The modern notation for the differential form, as well as the idea of the differential forms as being the wedge products of exterior derivatives forming an exterior algebra, was introduced by Élie Cartan. We initially work in an open set in R^n. A 0-form is defined to be a smooth function f. When we integrate a function f over an m-dimensional subspace S of R^n, we write it as

$$\int_S f\,dx^1 \cdots dx^m.$$

(The superscripts are indices, not exponents.) We can consider dx^1 through dx^n to be formal objects themselves, rather than tags appended to make integrals look like Riemann sums. Alternatively, we can view them as covectors, and thus a measure of "density" (hence integrable in a general sense). We call the $dx^1, \ldots, dx^n$ *basic* 1-*forms*.

We define the wedge product, "$\wedge$", a bilinear "multiplication" operator on these elements, with the *alternating* property that

$$dx^a \wedge dx^a = 0$$

for all indices a. Note that alternation along with linearity and associativity implies dx^{b}'–dx^{a} = –dx^{a}'–dx^{b}. This also ensures that the result of the wedge product has an orientation.

We define the set of all these products to be *basic 2-forms*, and similarly we define the set of products of the form dx^{a}'–dx^{b}'–dx^{c} to be *basic 3-forms*. A general k-form is then a weighted sum of basic k-forms, where the weights are the smooth functions f. Together these form a vector space with basic k-forms as the basis vectors, and 0-forms (smooth functions) as the field of scalars. The wedge product then extends to k-forms in the natural way. Over R^n at most n covectors can be linearly independent, thus a k-form with $k > n$ will always be zero, by the alternating property.

In addition to the wedge product, there is also the exterior derivative operator d. This operator maps k-forms to $(k+1)$-forms. For a k-form $\omega = f\, dx^a$ over R^n, we define the action of d by:

$$d\omega = \sum_{i=1}^{n} \frac{\partial f}{\partial x_i} dx^i \wedge dx^a.$$

with extension to general k-forms occurring linearly. This more general approach allows for a more natural coordinate-free approach to integration on manifolds. It also allows for a natural generalisation of the fundamental theorem of calculus, called Stokes' theorem, which we may state as

$$\int_{\Omega} d\omega = \int_{\partial\Omega} \omega$$

where ω is a general k-form, and $\partial\,\Omega$ denotes the boundary of the region Ω. Thus, in the case that ω is a 0-form and Ω is a closed interval of the real line, this reduces to the fundamental theorem of calculus. In the case that ω is a 1-form and Ω is a two-dimensional region in the plane, the theorem reduces to Green's theorem. Similarly, using 2-forms, and 3-forms and Hodge duality, we can arrive at Stokes' theorem and the divergence theorem.

Summations

The discrete equivalent of integration is summation. Summations and integrals can be put on the same foundations using the theory of Lebesgue integrals or time scale calculus.

Methods

Computing Integrals

The most basic technique for computing definite integrals of one real variable is based on the fundamental theorem of calculus. Let *f*(*x*) be the function of *x* to be integrated over a given interval [*a*, *b*]. Then, find an antiderivative of *f*; that is, a function *F* such that *F*' = *f* on the interval. Provided the integrand and integral have no singularities on the path of integration, by the fundamental theorem of calculus, $\int_a^b f(x)dx = F(b) - F(a)$.

The integral is not actually the antiderivative, but the fundamental theorem provides a way to use antiderivatives to evaluate definite integrals. The most difficult step is usually to find the antiderivative of *f*. It is rarely possible to glance at a function and write down its antiderivative. More often, it is necessary to use one of the many techniques that have been developed to evaluate integrals. Most of these techniques rewrite one integral as a different one which is hopefully more tractable. Techniques include:

- Integration by substitution
- Integration by parts
- Changing the order of integration
- Integration by trigonometric substitution
- Integration by partial fractions
- Integration by reduction formulae
- Integration using parametric derivatives
- Integration using Euler's formula
- Differentiation under the integral sign
- Contour integration

Alternate methods exist to compute more complex integrals. Many nonelementary integrals can be expanded in a Taylor series and integrated term by term. Occasionally, the resulting infinite series can be summed analytically. The method of convolution using Meijer G-functions can also be used, assuming that the integrand can be written as a product of Meijer G-functions. There are also many less common ways of calculating definite integrals; for instance, Parseval's identity can be used to transform an integral over a rectangular region into an infinite sum. Occasionally, an integral can be evaluated by a trick; for an example of this.

Computations of volumes of solids of revolution can usually be done with disk integration or shell integration. Specific results which have been worked out by various techniques are collected in the list of integrals.

Symbolic Algorithms

Many problems in mathematics, physics, and engineering involve integration where an explicit formula for the integral is desired. Extensive tables of integrals have been compiled and published over the years for this purpose. With the spread of computers, many professionals, educators, and students have turned to computer algebra systems that are specifically designed to perform difficult or tedious tasks, including integration. Symbolic integration has been one of the motivations for the development of the first such systems, like Macsyma.

A major mathematical difficulty in symbolic integration is that in many cases, a closed formula for the antiderivative of a rather simple-looking function does not exist. For instance, it is known that the antiderivatives of the functions $\exp(x^2)$, x^x and $(\sin x)/x$ cannot be expressed in the closed form involving only rational and exponential functions, logarithm, trigonometric and inverse trigonometric functions, and the operations of multiplication and composition; in other words, none of the three given functions is integrable in elementary functions, which are the functions which may be built from rational functions, roots of a polynomial, logarithm, and exponential functions. The Risch algorithm provides a general criterion to determine whether the antiderivative of an elementary function is elementary, and, if it is, to compute it.

Unfortunately, it turns out that functions with closed expressions of antiderivatives are the exception rather than the rule. Consequently, computerised algebra systems have no hope of being able to find an antiderivative for a randomly constructed elementary function. On the positive side, if the 'building blocks' for antiderivatives are fixed in advance, it may be still be possible to decide whether the antiderivative of a given function can be expressed using these blocks and operations of multiplication and composition, and to find the symbolic answer whenever it exists. The Risch algorithm, implemented in Mathematica and other computer algebra systems, does just that for functions and antiderivatives built from rational functions, radicals, logarithm, and exponential functions.

Some special integrands occur often enough to warrant special study. In particular, it may be useful to have, in the set of antiderivatives, the special functions of physics (like the Legendre functions, the hypergeometric function, the Gamma function, the Incomplete Gamma function and so on). Extending the Risch's algorithm to include such functions is possible but challenging and has been an active research subject.

More recently a new approach has emerged, using D-finite function, which are the solutions of linear differential equations with polynomial coefficients. Most of the elementary and special functions are D-finite and the integral of a D-finite function is also a D-finite function. This provide an algorithm to express the antiderivative of a D-finite function as the solution of a differential equation. This theory allows also to compute a definite integrals of a D-function as the sum of a series given by the first coefficients and an algorithm to compute any coefficient.

Numerical Quadrature

The integrals encountered in a basic calculus course are deliberately chosen for simplicity; those found in real applications are not always so accommodating. Some integrals cannot be found exactly, some require special functions which themselves are a challenge to compute, and others are so complex that finding the exact answer is too slow.

This motivates the study and application of numerical methods for approximating integrals, which today use floating-point arithmetic on digital electronic computers. Many of the ideas arose much earlier, for hand calculations; but the speed of general-purpose computers like the ENIAC created a need for improvements.

The goals of numerical integration are accuracy, reliability, efficiency, and generality. Sophisticated methods can vastly outperform a naive method by all four measures (Dahlquist & Björck 2008; Kahaner, Moler & Nash 1989; Stoer & Bulirsch 2002). Consider, for example, the integral

$$\int_{-2}^{2} \frac{1}{5}\left(\frac{1}{100}(322 + 3x(98 + x(37 + x))) - 24\frac{x}{1+x^2} \right) dx,$$

which has the exact answer 94/25 = 3.76. (In ordinary practice the answer is not known in advance, so an important task — not explored here — is to decide when an approximation is good enough.) A "calculus book" approach divides the integration range into, say, 16 equal pieces, and computes function values.

Spaced function values									
x	?2.00	?1.50	?1.00	?0.50	0.00	0.50	1.00	1.50	2.00
f(x)	2.22800	2.45663	2.67200	2.32475	0.64400	?0.92575	?0.94000	?0.16963	0.83600
x	?1.75	?1.25	?0.75	?0.25	0.25	0.75	1.25	1.75	
f(x)	2.33041	2.58562	2.62934	1.64019	?0.32444	?1.09159	?0.60387	0.31734	

Using the left end of each piece, the rectangle method sums 16 function values and multiplies by the step width, h, here 0.25, to get an approximate value of 3.94325 for the integral. The accuracy is not impressive, but calculus formally uses pieces of infinitesimal width, so initially this may seem little cause for concern. Indeed, repeatedly doubling the number of steps eventually produces an approximation of 3.76001. However, 2^{18} pieces are required, a great computational expense for such little accuracy; and a reach for greater accuracy can force steps so small that arithmetic precision becomes an obstacle.

A better approach replaces the horizontal tops of the rectangles with slanted tops touching the function at the ends of each piece. This trapezium rule is almost as easy to calculate; it sums all 17 function values, but weights the first and last by one half, and again multiplies by the step width. This immediately improves the approximation to 3.76925, which is noticeably more accurate. Furthermore, only 2^{10} pieces are needed to achieve 3.76000, substantially less computation than the rectangle method for comparable accuracy.

Romberg's method builds on the trapezoid method to great effect. First, the step lengths are halved incrementally, giving trapezoid approximations denoted by $T(h_0)$, $T(h_1)$, and so on, where h_{k+1} is half of h_k. For each new step size, only half the new function values need to be computed; the others carry over from the previous size. But the really powerful idea is to interpolate a polynomial through the approximations, and extrapolate to $T(0)$. With this method a numerically *exact* answer here requires only four pieces (five function values)! The Lagrange polynomial interpolating $\{h_k, T(h_k)\}_{k=0...2} = \{(4.00,6.128), (2.00,4.352), (1.00,3.908)\}$ is $3.76 + 0.148h^2$, producing the extrapolated value 3.76 at $h = 0$.

Gaussian quadrature often requires noticeably less work for superior accuracy. In this example, it can compute the function values at just two x positions, ±2D −3, then double each value and sum to get the numerically exact answer. The explanation for this dramatic success lies in error analysis, and a little luck. An n-point Gaussian method is exact for polynomials of degree up to $2n-1$. The function in this example is a degree 3 polynomial, plus a term that cancels because the chosen endpoints are symmetric around zero. (Cancellation also benefits the Romberg method.)

Shifting the range left a little, so the integral is from −2.25 to 1.75, removes the symmetry. Nevertheless, the trapezoid method is rather slow, the polynomial interpolation method of Romberg is acceptable, and the Gaussian method requires the least work — if the number of points is known in advance. As well, rational interpolation can use the same trapezoid evaluations as the Romberg method to greater effect.

Quadrature method cost comparison				
Method	*Trapezoid*	*Romberg*	*Rational*	*Gauss*
Points	1048577	257	129	36
Rel. Err.	-5.3×10^{-13}	-6.3×10^{-15}	8.8×10^{-15}	3.1×10^{-15}
Value	$\int_{-2.25}^{1.75} f(x)dx = 4.1639019006585897075\ldots$			

In practice, each method must use extra evaluations to ensure an error bound on an unknown function; this tends to offset some of the advantage of the pure Gaussian method, and motivates the popular Gauss–Kronrod quadrature formulae. Symmetry can still be exploited by splitting this integral into two ranges, from −2.25 to −1.75 (no symmetry), and from −1.75 to 1.75 (symmetry). More broadly, adaptive quadrature partitions a range into pieces based on function properties, so that data points are concentrated where they are needed most.

Simpson's rule, named for Thomas Simpson (1710–1761), uses a parabolic curve to approximate integrals. In many cases, it is more accurate than the trapezoidal rule and others. The rule states that

$$\int_a^b f(x)dx \approx \frac{b-a}{6}\left[f(a)+4f\left(\frac{a+b}{2}\right)+f(b)\right],$$

with an error of

$$\left|-\frac{(b-a)^5}{2880}f^{(4)}(\xi)\right|.$$

The computation of higher-dimensional integrals (for example, volume calculations) makes important use of such alternatives as Monte Carlo integration.

A calculus text is no substitute for numerical analysis, but the reverse is also true. Even the best adaptive numerical code sometimes requires a user to help with the more demanding integrals. For example, improper integrals may require a change of variable or methods that can avoid infinite function values, and known properties like symmetry and periodicity may provide critical leverage.

Numerical Integration

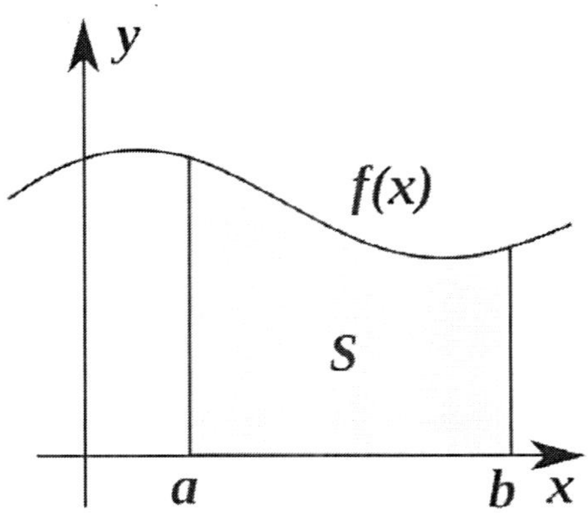

Figure: *Numerical integration consists of finding numerical approximations for the value* S

In numerical analysis, numerical integration constitutes a broad family of algorithms for calculating the numerical value of a definite integral, and by extension, the term is also sometimes used to describe the numerical solution of differential equations. This article focuses on calculation of definite integrals. The term numerical quadrature (often abbreviated to *quadrature*) is more or less a synonym for *numerical integration*, especially as applied to one-dimensional integrals. Numerical integration over more than one dimension is sometimes described as cubature, although the meaning of *quadrature* is understood for higher dimensional integration as well.

The basic problem considered by numerical integration is to compute an approximate solution to a definite integral:

$$\int_a^b f(x)dx.$$

If $f(x)$ is a smooth well-behaved function, integrated over a small number of dimensions and the limits of integration are bounded, there are many methods of approximating the integral with arbitrary precision.

Reasons for Numerical Integration

There are several reasons for carrying out numerical integration. The integrand $f(x)$ may be known only at certain points, such as obtained by sampling. Some embedded systems and other computer applications may need numerical integration for this reason.

A formula for the integrand may be known, but it may be difficult or impossible to find an antiderivative which is an elementary function. An example of such an integrand is $f(x) = \exp(-x^2)$, the antiderivative of which (the error function, times a constant) cannot be written in elementary form.

It may be possible to find an antiderivative symbolically, but it may be easier to compute a numerical approximation than to compute the antiderivative. That may be the case if the antiderivative is given as an infinite series or product, or if its evaluation requires a special function which is not available.

Methods for One-dimensional Integrals

Numerical integration methods can generally be described as combining evaluations of the integrand to get an approximation to the integral. The integrand is evaluated at a finite set of points called integration points and a weighted sum of these values is used to approximate the integral. The integration points and weights depend on the specific method used and the accuracy required from the approximation.

An important part of the analysis of any numerical integration method is to study the behaviour of the approximation error as a function of the number of integrand evaluations. A method which yields a small error for a small number of evaluations is usually considered superior. Reducing the number of evaluations of the integrand reduces the number of arithmetic operations involved, and therefore reduces the total round-off error. Also, each evaluation takes time, and the integrand may be arbitrarily complicated.

A 'brute force' kind of numerical integration can be done, if the integrand is reasonably well-behaved (i.e. piecewise continuous and of bounded variation), by evaluating the integrand with very small increments.

Quadrature Rules Based on Interpolating Functions

A large class of quadrature rules can be derived by constructing interpolating functions which are easy to integrate. Typically these interpolating functions are polynomials.

The simplest method of this type is to let the interpolating function be a constant function (a polynomial of degree zero) which passes through the point $((a+b)/2, f((a+b)/2))$. This is called the *midpoint rule* or *rectangle rule.*

$$\int_a^b f(x)dx \approx (b-a)f\left(\frac{a+b}{2}\right).$$

The interpolating function may be an affine function (a polynomial of degree 1) which passes through the points $(a, f(a))$ and $(b, f(b))$. This is called the *trapezoidal rule.*

$$\int_a^b f(x)dx \approx (b-a)\frac{f(a)+f(b)}{2}.$$

For either one of these rules, we can make a more accurate approximation by breaking up the interval $[a, b]$ into some number n of subintervals, computing an approximation for each subinterval, then adding up all the results. This is called a *composite rule*, *extended rule*, or *iterated rule*. For example, the composite trapezoidal rule can be stated as

$$\int_a^b f(x)dx \approx \frac{b-a}{n}\left(\frac{f(a)+f(b)}{2}+\sum_{k=1}^{n-1} f\left(a+k\frac{b-a}{n}\right)\right)$$

where the subintervals have the form $[k\ h, (k+1)\ h]$, with $h = (b-a)/n$ and $k = 0, 1, 2,..., n-1$.

Interpolation with polynomials evaluated at equally spaced points in $[a, b]$ yields the Newton–Cotes formulas, of which the rectangle rule and the trapezoidal rule are examples. Simpson's rule, which is based on a polynomial of order 2, is also a Newton–Cotes formula.

Quadrature rules with equally spaced points have the very convenient property of *nesting*. The corresponding rule with each interval subdivided includes all the current points, so those integrand values can be re-used.

If we allow the intervals between interpolation points to vary, we find another group of quadrature formulas, such as the Gaussian quadrature formulas. A Gaussian quadrature rule is typically more accurate than a Newton–Cotes rule which requires the same number of function evaluations, if the integrand is smooth (i.e., if it is sufficiently differentiable). Other quadrature methods with varying intervals include Clenshaw–Curtis quadrature (also called Fejér quadrature) methods, which do nest.

Gaussian quadrature rules do not nest, but the related Gauss–Kronrod quadrature formulas do.

Adaptive Algorithms

If *f(x)* does not have many derivatives at all points, or if the derivatives become large, then Gaussian quadrature is often insufficient. In this case, an algorithm similar to the following will perform better:

```
def calculate_definite_integral_of_f(f, initial_step_size):
    '''
    This algorithm calculates the definite integral
    of a function from 0 to 1, adaptively, by choosing
    smaller steps near problematic points.
```

```
'''
x = 0.0
h = initial_step_size
accumulator = 0.0
while x < 1.0:
    if x + h > 1.0:
        h = 1.0 - x
    quad_this_step =
    if error_too_big_in_quadrature_of_over_range(f,
    [x,x+h]):
    h = make_h_smaller(h)
    else:
    accumulator += quadrature_of_f_over_range(f,
    [x,x+h])
    x += h
    if error_too_small_in_quadrature_of_over_range(f,
    [x,x+h]):
    h = make_h_larger(h) # Avoid wasting time on
    tiny steps.
return accumulator
```

Some details of the algorithm require careful thought. For many cases, estimating the error from quadrature over an interval for a function $f(x)$ isn't obvious. One popular solution is to use two different rules of quadrature, and use their difference as an estimate of the error from quadrature.

The other problem is deciding what "too large" or "very small" signify. A *local* criterion for "too large" is that the quadrature error should not be larger than $t \cdot h$ where t, a real number, is the tolerance we wish to set for global error.

Then again, if h is already tiny, it may not be worthwhile to make it even smaller even if the quadrature error is apparently large. A *global* criterion is that the sum of errors on all the intervals should be less than t. This type of error analysis is usually called "a posteriori" since we compute the error after having computed the approximation.

Heuristics for adaptive quadrature are discussed by Forsythe et al. (Section 5.4).

Extrapolation Methods

The accuracy of a quadrature rule of the Newton-Cotes type is generally a function of the number of evaluation points. The result is usually more accurate as number of evaluation points increases, or, equivalently, as the width of the step size between the points decreases.

It is natural to ask what the result would be if the step size were allowed to approach zero. This can be answered by extrapolating the result from two or more nonzero step sizes, using series acceleration methods such as Richardson extrapolation.

The extrapolation function may be a polynomial or rational function. Extrapolation methods are described in more detail by Stoer and Bulirsch (Section 3.4) and are implemented in many of the routines in the QUADPACK library.

Conservative (a priori) Error Estimation

Let f have a bounded first derivative over $[a,b]$. The mean value theorem for f, where $x < b$, gives

$$(x-a)f'(y_x) = f(x) - f(a)$$

for some y_x in $[a,x]$ depending on x. If we integrate in x from a to b on both sides and take the absolute values, we obtain

$$\left|\int_a^b f(x)dx - (b-a)f(a)\right| = \left|\int_a^b (x-a)f'(y_x)dx\right|$$

We can further approximate the integral on the right-hand side by bringing the absolute value into the integrand, and replacing the term in f' by an upper bound:

$$\left|\int_a^b f(x)dx - (b-a)f(a)\right| \le \frac{(b-a)^2}{2}\sup_{a\le x\le b}\left|f'(x)\right|$$

Hence, if we approximate the integral +–$_a^b$ $f(x)$ dx by the quadrature rule $(b-a)f(a)$ our error is no greater than the right hand side of (**). We can convert this into an error analysis for the Riemann sum (*), giving an upper bound of

$$\frac{n^{-1}}{2}\sup_{0\le x\le 1}\left|f'(x)\right|$$

for the error term of that particular approximation. (Note that this is precisely the error we calculated for the example $f(x) = x$.)

Using more derivatives, and by tweaking the quadrature, we can do a similar error analysis using a Taylor series (using a partial sum with remainder term) for f.

This error analysis gives a strict upper bound on the error, if the derivatives of f are available.

This integration method can be combined with interval arithmetic to produce computer proofs and *verified* calculations.

Integrals Over Infinite Intervals

Infinite Intervals

One way to calculate an integral over infinite interval,

$$\int_{-\infty}^{+\infty} f(x)dx,$$

is to transform it into an integral over a finite interval by any one of several possible changes of variables, for example:

$$\int_{-\infty}^{+\infty} f(x)dx = \int_{-1}^{+1} f\left(\frac{t}{1-t^2}\right)\frac{1+t^2}{(1-t^2)^2}dt,$$

The integral over finite interval can then be evaluated by ordinary integration methods.

Half-infinite intervals

An integral over a half-infinite interval can likewise be transformed into an integral over a finite interval by any one of several possible changes of variables, for example:

$$\int_{a}^{+\infty} f(x)dx = \int_{0}^{1} f\left(a+\frac{1-t}{t}\right)\frac{dt}{t^2}.$$

Similarly,

$$\int_{-\infty}^{a} f(x)dx = \int_{0}^{1} f\left(a-\frac{1-t}{t}\right)\frac{dt}{t^2}$$

Multidimensional Integrals

The quadrature rules discussed so far are all designed to compute one-dimensional integrals.

To compute integrals in multiple dimensions, one approach is to phrase the multiple integral as repeated one-dimensional integrals by appealing to Fubini's theorem.

This approach requires the function evaluations to grow exponentially as the number of dimensions increases. Two methods are known to overcome this so-called *curse of dimensionality*.

Monte Carlo

Monte Carlo methods and quasi-Monte Carlo methods are easy to apply to multi-dimensional integrals, and may yield greater accuracy for the same number of function evaluations than repeated integrations using one-dimensional methods.

A large class of useful Monte Carlo methods are the so-called Markov chain Monte Carlo algorithms, which include the Metropolis-Hastings algorithm and Gibbs sampling.

Sparse Grids

Sparse grids were originally developed by Smolyak for the quadrature of high dimensional functions.

The method is always based on a one dimensional quadrature rule, but performs a more sophisticated combination of univariate results.

Connection with Differential Equations

The problem of evaluating the integral

$$\int_a^b f(x)dx$$

can be reduced to an initial value problem for an ordinary differential equation. If the above integral is denoted by $I(b)$, then the function I satisfies

$$I'(x) = f(x), \quad I(a) = 0.$$

Methods developed for ordinary differential equations, such as Runge–Kutta methods, can be applied to the restated problem and thus be used to evaluate the integral. For instance, the standard fourth-order Runge–Kutta method applied to the differential equation yields Simpson's rule from above.

The differential equation $I'(x) = f(x)$ has a special form: the right-hand side contains only the dependent variable (here x) and not the independent variable (here I). This simplifies the theory and algorithms considerably. The problem of evaluating integrals is thus best studied in its own right.

Dirac Delta Function

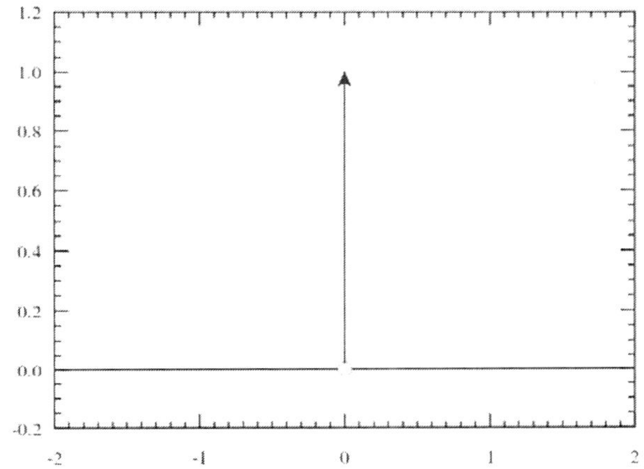

Figure: *Schematic representation of the Dirac delta function by a line surmounted by an arrow. The height of the arrow is usually used to specify the value of any multiplicative constant, which will give the area under the function. The other convention is to write the area next to the arrowhead.*

The Dirac delta function, or δ function, is (informally) a generalised function on the real number line that is zero everywhere except at zero, with an integral of one over the entire real line. The delta function is sometimes thought of as an infinitely high, infinitely thin spike at the origin, with total area one under the spike, and physically represents an idealised point mass or point charge. It was introduced by theoretical physicist Paul Dirac. Dirac explicitly spoke of infinitely great values of his integrand. In the context of signal processing it is often referred to as the unit impulse symbol (or function). Its discrete analog is the Kronecker delta function which is usually defined on a finite domain and takes values 0 and 1.

From a purely mathematical viewpoint, the Dirac delta is not strictly a function, because any extended-real function that is equal to zero everywhere but a single point must have total integral zero. The delta function only makes sense as a mathematical object when it appears inside an integral. While from this perspective the Dirac delta can usually be manipulated as though it were a function, formally it must be defined as a distribution that is also a measure. In many applications, the

Dirac delta is regarded as a kind of limit (a weak limit) of a sequence of functions having a tall spike at the origin. The approximating functions of the sequence are thus "approximate" or "nascent" delta functions.

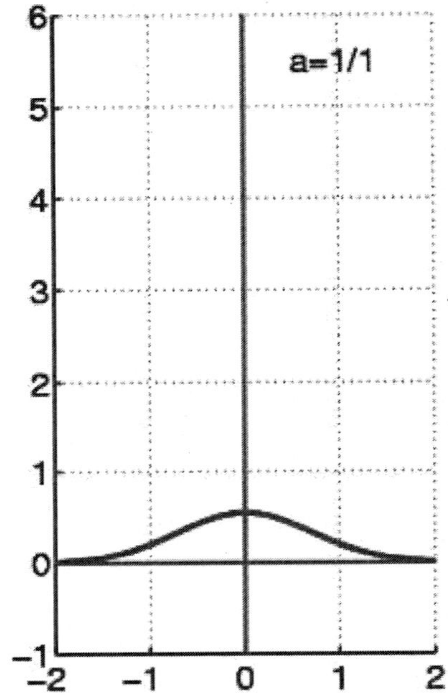

***Figure:** The Dirac delta function as the limit (in the sense of distributions) of the sequence of zero-centred normal distributions*

$$\delta_a(x) = \frac{1}{a\sqrt{\pi}} e^{-x^2/a^2} \quad as \ a \to 0.$$

Overview

The graph of the delta function is usually thought of as following the whole x-axis and the positive y-axis. Despite its name, the delta function is not truly a function, at least not a usual one with domain in reals. For example, the objects $f(x) = \delta(x)$ and $g(x) = 0$ are equal everywhere except at $x = 0$ yet have integrals that are different.

According to Lebesgue integration theory, if f and g are functions such that $f = g$ almost everywhere, then f is integrable if and only if g is integrable and the integrals of f and g are identical. Rigorous treatment of the Dirac delta requires measure theory or the theory of distributions.

The Dirac delta is used to model a tall narrow spike function (an *impulse*), and other similar abstractions such as a point charge, point mass or electron point. For example, to calculate the dynamics of a baseball being hit by a bat, one can approximate the force of the bat hitting the baseball by a delta function.

In doing so, one not only simplifies the equations, but one also is able to calculate the motion of the baseball by only considering the total impulse of the bat against the ball rather than requiring knowledge of the details of how the bat transferred energy to the ball.

In applied mathematics, the delta function is often manipulated as a kind of limit (a weak limit) of a sequence of functions, each member of which has a tall spike at the origin: for example, a sequence of Gaussian distributions centred at the origin with variance tending to zero.

History

Joseph Fourier presented what is now called the *Fourier integral theorem* in his treatise *Théorie analytique de la chaleur* in the form:

$$f(x) = \frac{1}{2\pi}\int_{-\infty}^{\infty} d\alpha f(\alpha) \int_{-\infty}^{\infty} dp \cos(px - p\alpha),$$

which is tantamount to the introduction of the δ-function in the form:

$$\delta(x-\alpha) = \frac{1}{2\pi}\int_{-\infty}^{\infty} dp \cos(px - p\alpha).$$

Later, Augustin Cauchy expressed the theorem using exponentials:

$$f(x) = \frac{1}{2\pi}\int_{-\infty}^{\infty} e^{ipx}\left(\int_{-\infty}^{\infty} e^{-ip\alpha} f(\alpha)\, d\alpha\right) dp.$$

Cauchy pointed out that in some circumstances the *order* of integration in this result was significant. As justified using the theory of distributions, the Cauchy equation can be rearranged to resemble Fourier's original formulation and expose the δ-function as:

$$f(x) = \frac{1}{2\pi}\int_{-\infty}^{\infty} e^{ipx}\left(\int_{-\infty}^{\infty} e^{-ip\alpha} f(\alpha)\, d\alpha\right) dp$$

$$= \frac{1}{2\pi}\int_{-\infty}^{\infty}\left(\int_{-\infty}^{\infty} e^{ipx}e^{-ip\alpha}\, dp\right) f(\alpha)\, d\alpha = \int_{-\infty}^{\infty} \delta(x-\alpha) f(\alpha)\, d\alpha\,,$$

where the δ-function is expressed as:

$$\delta(x-\alpha) = \frac{1}{2\pi}\int_{-\infty}^{\infty} e^{ipx}e^{-ip\alpha}\, dp\,.$$

A rigorous interpretation of the exponential form and the various limitations upon the function f necessary for its application extended over several centuries. The problems with a classical interpretation are explained as follows:

> *"The greatest drawback of the classical Fourier transformation is a rather narrow class of functions (originals) for which it can be effectively computed. Namely, it is necessary that these functions decrease sufficiently rapidly to zero (in the neighbourhood of infinity) in order to insure the existence of the Fourier integral. For example, the Fourier transform of such simple functions as polynomials does not exist in the classical sense. The extension of the classical Fourier transformation to distributions considerably enlarged the class of functions that could be transformed and this removed many obstacles."*

Further developments included generalisation of the Fourier integral, "beginning with Plancherel's pathbreaking L^2-theory (1910), continuing with Wiener's and Bochner's works (around 1930) and culminating with the amalgamation into L. Schwartz's theory of distributions (1945)...", and leading to the formal development of the Dirac delta function.

An infinitesimal formula for an infinitely tall, unit impulse delta function (infinitesimal version of Cauchy distribution) explicitly appears in an 1827 text of Augustin Louis Cauchy. Siméon Denis Poisson considered the issue in connection with the study of wave propagation as did Gustav Kirchhoff somewhat later. Kirchhoff and Hermann von Helmholtz also introduced the unit impulse as a limit of Gaussians, which also corresponded to Lord Kelvin's notion of a point heat source.

At the end of the 19th century, Oliver Heaviside used formal Fourier series to manipulate the unit impulse. The Dirac delta function as such was introduced as a "convenient notation" by Paul Dirac in his influential 1930 book *Principles of Quantum Mechanics*. He called it the "delta function" since he used it as a continuous analogue of the discrete Kronecker delta.

Definitions

The Dirac delta can be loosely thought of as a function on the real line which is zero everywhere except at the origin, where it is infinite,

$$\delta(x) = \begin{cases} +\infty, & x = 0 \\ 0, & x \neq 0 \end{cases}$$

and which is also constrained to satisfy the identity

$$\int_{-\infty}^{\infty} \delta(x) dx = 1.$$

This is merely a heuristic characterisation. The Dirac delta is not a function in the traditional sense as no function defined on the real numbers has these properties. The Dirac delta function can be rigorously defined either as a distribution or as a measure.

As a Measure

One way to rigorously define the delta function is as a measure, which accepts as an argument a subset A of the real line R, and returns $\delta(A) = 1$ if $0 - A$, and $\delta(A) = 0$ otherwise. If the delta function is conceptualised as modelling an idealised point mass at 0, then $\delta(A)$ represents the mass contained in the set A. One may then define the integral against δ as the integral of a function against this mass distribution. Formally, the Lebesgue integral provides the necessary analytic device. The Lebesgue integral with respect to the measure δ satisfies

$$\int_{-\infty}^{\infty} f(x)\delta\{dx\} = f(0)$$

for all continuous compactly supported functions f. The measure δ is not absolutely continuous with respect to the Lebesgue measure — in fact, it is a singular measure. Consequently, the delta measure has no Radon–Nikodym derivative — no true function for which the property

$$\int_{-\infty}^{\infty} f(x)\delta(x)dx = f(0)$$

holds. As a result, the latter notation is a convenient abuse of notation, and not a standard (Riemann or Lebesgue) integral.

As a probability measure on R, the delta measure is characterised by its cumulative distribution function, which is the unit step function

$$H(x) = \begin{cases} 1 & \text{if } x \geq 0 \\ 0 & \text{if } x < 0. \end{cases}$$

This means that $H(x)$ is the integral of the cumulative indicator function $1_{(\in -, x]}$ with respect to the measure δ; to wit,

$$H(x) = \int_{\mathbb{R}} \mathbf{1}_{(-\infty,x]}(t)\delta\{dt\} = \delta(-\infty, x].$$

Thus in particular the integral of the delta function against a continuous function can be properly understood as a Stieltjes integral:

$$\int_{-\infty}^{\infty} f(x)\delta\{dx\} = \int_{-\infty}^{\infty} f(x)dH(x).$$

All higher moments of δ are zero. In particular, characteristic function and moment generating function are both equal to one.

As a Distribution

In the theory of distributions a generalised function is thought of not as a function itself, but only in relation to how it affects other functions when it is "integrated" against them. In keeping with this philosophy, to define the delta function properly, it is enough to say what the "integral" of the delta function against a sufficiently "good" test function is.

If the delta function is already understood as a measure, then the Lebesgue integral of a test function against that measure supplies the necessary integral. A typical space of test functions consists of all smooth functions on R with compact support. As a distribution, the Dirac delta is a linear functional on the space of test functions and is defined by

$$\delta[\varphi] = \varphi(0) \tag{1}$$

for every test function φ.

For δ to be properly a distribution, it must be "continuous" in a suitable sense. In general, for a linear functional S on the space of test functions to define a distribution, it is necessary and sufficient that, for every positive integer N there is an integer M_N and a constant C_N such that for every test function φ, one has the inequality

$$|S[\phi]| \leq C_N \sum_{k=0}^{M_N} \sup_{x\in[-N,N]} |\phi^{(k)}(x)|.$$

With the δ distribution, one has such an inequality (with $C_N = 1$) with $M_N = 0$ for all N. Thus δ is a distribution of order zero. It is, furthermore, a distribution with compact support (the support being $\{0\}$).

The delta distribution can also be defined in a number of equivalent ways. For instance, it is the distributional derivative of the Heaviside step function. This means that, for every test function φ, one has

$$\delta[\phi] = -\int_{-\infty}^{\infty} \phi'(x)H(x)dx.$$

Intuitively, if integration by parts were permitted, then the latter integral should simplify to

$$\int_{-\infty}^{\infty} \phi(x)H'(x)dx = \int_{-\infty}^{\infty} \phi(x)\delta(x)dx,$$

and indeed, a form of integration by parts is permitted for the Stieltjes integral, and in that case one does have

$$-\int_{-\infty}^{\infty} \phi'(x)H(x)dx = \int_{-\infty}^{\infty} \phi(x)dH(x).$$

In the context of measure theory, the Dirac measure gives rise to a distribution by integration. Conversely, equation (1) defines a Daniell integral on the space of all compactly supported continuous functions φ which, by the Riesz representation theorem, can be represented as the Lebesgue integral of φ with respect to some Radon measure.

Generalisations

The delta function can be defined in n-dimensional Euclidean space R^n as the measure such that

$$\int_{\mathbf{R}^n} f(\mathbf{x})\delta\{d\mathbf{x}\} = f(\mathbf{0})$$

for every compactly supported continuous function f. As a measure, the n-dimensional delta function is the product measure of the 1-dimensional delta functions in each variable separately. Thus, formally, with $x = (x_1, x_2, \ldots, x_n)$, one has

$$\delta(\mathbf{x}) = \delta(x_1)\delta(x_2)\ldots\delta(x_n). \qquad (2)$$

The delta function can also be defined in the sense of distributions exactly as above in the one-dimensional case. However, despite widespread use in engineering contexts, (2) should be manipulated with care, since the product of distributions can only be defined under quite narrow circumstances.

The notion of a Dirac measure makes sense on any set whatsoever. Thus if X is a set, $x_0 \in X$ is a marked point, and Σ is any sigma algebra of subsets of X, then the measure defined on sets $A \in \Sigma$ by

$$\delta_{x_0}(A) = \begin{cases} 1 & \text{if } x_0 \in A \\ 0 & \text{if } x_0 \notin A \end{cases}$$

is the delta measure or unit mass concentrated at x_0.

Another common generalisation of the delta function is to a differentiable manifold where most of its properties as a distribution can also be exploited because of the differentiable structure. The delta function on a manifold M centred at the point $x_0 - M$ is defined as the following distribution:

$$\delta_{x_0}[\phi] = \phi(x_0) \qquad (3)$$

for all compactly supported smooth real-valued functions φ on M. A common special case of this construction is when M is an open set in the Euclidean space R^n.

On a locally compact Hausdorff space X, the Dirac delta measure concentrated at a point x is the Radon measure associated with the Daniell integral (3) on compactly supported continuous functions φ. At this level of generality, calculus as such is no longer possible, however a variety of techniques from abstract analysis are available.

For instance, the mapping $x_0 \mapsto \delta_{x_0}$ is a continuous embedding of X into the space of finite Radon measures on X, equipped with its vague topology. Moreover, the convex hull of the image of X under this embedding is dense in the space of probability measures on X.Properties

Scaling and Symmetry

The delta function satisfies the following scaling property for a non-zero scalar α:

$$\int_{-\infty}^{\infty} \delta(\alpha x)dx = \int_{-\infty}^{\infty} \delta(u)\frac{du}{|\alpha|} = \frac{1}{|\alpha|}$$

and so

$$\delta(\alpha x) = \frac{\delta(x)}{|\alpha|}. \qquad (4)$$

In particular, the delta function is an even distribution, in the sense that

$$\delta(-x) = \delta(x)$$

which is homogeneous of degree −1.

Algebraic Properties

The distributional product of δ with x is equal to zero:

$$x\delta(x) = 0.$$

Conversely, if $xf(x) = xg(x)$, where f and g are distributions, then

$$f(x) = g(x) + c\delta(x)$$

for some constant c.

Translation

The integral of the time-delayed Dirac delta is given by:

$$\int_{-\infty}^{\infty} f(t)\delta(t-T)dt = f(T)$$

This is sometimes referred to as the *sifting property* or the *sampling property*. The delta function is said to "sift out" the value at $t = T$.

It follows that the effect of convolving a function $f(t)$ with the time-delayed Dirac delta is to time-delay $f(t)$ by the same amount:

$$(f(t) * \delta(t-T)) \overset{\text{def}}{=} \int_{-\infty}^{\infty} f(\tau)\cdot\delta(t-T-\tau)d\tau$$

$$= \int_{-\infty}^{\infty} f(\tau)\cdot\delta(\tau-(t-T))d\tau \quad (\text{using } (4)\text{: } \delta(-x)=\delta(x))$$

$$= f(t-T)$$

This holds under the precise condition that f be a tempered distribution. As a special case, for instance, we have the identity (understood in the distribution sense)

$$\int_{-\infty}^{\infty} \delta(\xi - x)\delta(x-\eta)dx = \delta(\xi-\eta).$$

Composition with a Function

More generally, the delta distribution may be composed with a smooth function $g(x)$ in such a way that the familiar change of variables formula holds, that

$$\int_{\mathbf{R}} \delta(g(x)) f(g(x)) \mid g'(x) \mid dx = \int_{g(\mathbf{R})} \delta(u) f(u) du$$

provided that g is a continuously differentiable function with $g2$ nowhere zero. That is, there is a unique way to assign meaning to the distribution $\delta^{\circ}g$ so that this identity holds for all compactly supported test functions f. This distribution satisfies $\delta(g(x)) = 0$ if g is nowhere zero, and otherwise if g has a real root at x_0, then

$$\delta(g(x)) = \frac{\delta(x-x_0)}{\mid g'(x_0) \mid}.$$

It is natural therefore to *define* the composition $\delta(g(x))$ for continuously differentiable functions g by

$$\delta(g(x)) = \sum_i \frac{\delta(x-x_i)}{\mid g'(x_i) \mid}$$

where the sum extends over all roots of $g(x)$, which are assumed to be simple. Thus, for example

$$\delta(x^2 - \alpha^2) = \frac{1}{2 \mid \alpha \mid}[\delta(x+\alpha) + \delta(x-\alpha)]$$

In the integral form the generalised scaling property may be written as

$$\int_{-\infty}^{\infty} f(x)\delta(g(x))dx = \sum_i \frac{f(x_i)}{|g'(x_i)|}$$

Properties in n Dimensions

The delta distribution in an n-dimensional space satisfies the following scaling property instead:

$$\delta(\alpha\mathbf{x}) = |\alpha|^{-n}\ \delta(\mathbf{x})$$

so that δ is a homogeneous distribution of degree $-n$. Under any reflection or rotation ρ, the delta function is invariant:

$$\delta(\rho\mathbf{x}) = \delta(\mathbf{x}).$$

As in the one-variable case, it is possible to define the composition of δ with a bi-Lipschitz function $g : \mathrm{R}^n \to \mathrm{R}^n$ uniquely so that the identity

$$\int_{\mathrm{R}^n} \delta(g(\mathbf{x}))\, f(g(\mathbf{x}))|\det g'(\mathbf{x})|d\mathbf{x} = \int_{g(\mathrm{R}^n)} \delta(\mathbf{u})f(\mathbf{u})d\mathbf{u}$$

for all compactly supported functions f.

Using the coarea formula from geometric measure theory, one can also define the composition of the delta function with a submersion from one Euclidean space to another one of different dimension; the result is a type of current. In the special case of a continuously differentiable function $g : \mathrm{R}^n \to \mathrm{R}$ such that the gradient of g is nowhere zero, the following identity holds

$$\int_{\mathbb{R}^n} f(\mathbf{x})\delta(g(\mathbf{x}))d\mathbf{x} = \int_{g^{-1}(0)} \frac{f(\mathbf{x})}{|\nabla g|} d\sigma(\mathbf{x})$$

where the integral on the right is over $g^{-1}(0)$, the $n-1$ dimensional surface defined by $g(\mathrm{x}) = 0$ with respect to the Minkowski content measure. This is known as a simple layer integral.

Fourier Transform

The delta function is a tempered distribution, and therefore it has a well-defined Fourier transform. Formally, one finds

$$\hat{\delta}(\xi) = \int_{-\infty}^{\infty} e^{-2\pi i x\xi}\delta(x)dx = 1.$$

Properly speaking, the Fourier transform of a distribution is defined by imposing self-adjointness of the Fourier transform under the duality pairing $\langle\cdot,\cdot\rangle$ of tempered distributions with Schwartz functions. Thus $\hat{\delta}$ is defined as the unique tempered distribution satisfying

$$\langle\hat{\delta},\phi\rangle = \langle\delta,\hat{\phi}\rangle$$

for all Schwartz functions φ. And indeed it follows from this that $\hat{\delta}=1$.

As a result of this identity, the convolution of the delta function with any other tempered distribution S is simply S:

$$S*\delta = S.$$

That is to say that δ is an identity element for the convolution on tempered distributions, and in fact the space of compactly supported distributions under convolution is an associative algebra with identity the delta function. This property is fundamental in signal processing, as convolution with a tempered distribution is a linear time-invariant system, and applying the linear time-invariant system measures its impulse response.

The impulse response can be computed to any desired degree of accuracy by choosing a suitable approximation for δ, and once it is known, it characterises the system completely.

The inverse Fourier transform of the tempered distribution $f(\partial) = 1$ is the delta function. Formally, this is expressed

$$\int_{-\infty}^{\infty} 1\cdot e^{2\pi i x\xi}\,d\xi = \delta(x)$$

and more rigorously, it follows since

$$\langle 1, f^{\vee}\rangle = f(0) = \langle\delta, f\rangle$$

for all Schwartz functions f.

In these terms, the delta function provides a suggestive statement of the orthogonality property of the Fourier kernel on R. Formally, one has

$$\int_{-\infty}^{\infty} e^{i2\pi\xi_1 t}\left[e^{i2\pi\xi_2 t}\right]^* dt = \int_{-\infty}^{\infty} e^{-i2\pi(\xi_2-\xi_1)t}\,dt = \delta(\xi_1-\xi_2).$$

This is, of course, shorthand for the assertion that the Fourier transform of the tempered distribution

$$f(t) = e^{i2\pi\xi_1 t}$$

is

$$\hat{f}(\xi_2) = \delta(\xi_1 - \xi_2)$$

which again follows by imposing self-adjointness of the Fourier transform.

By analytic continuation of the Fourier transform, the Laplace transform of the delta function is found to be

$$\int_0^\infty \delta(t-a)e^{-st}\,dt = e^{-sa}.$$

Distributional Derivatives

The distributional derivative of the Dirac delta distribution is the distribution δ2 defined on compactly supported smooth test functions φ by

$$\delta'[\varphi] = -\delta[\varphi'] = -\varphi'(0).$$

The first equality here is a kind of integration by parts, for if δ were a true function then

$$\int_{-\infty}^{\infty} \delta'(x)\varphi(x)dx = -\int_{-\infty}^{\infty} \delta(x)\varphi'(x)dx.$$

The k-th derivative of δ is defined similarly as the distribution given on test functions by

$$\delta^{(k)}[\varphi] = (-1)^k \varphi^{(k)}(0).$$

In particular δ is an infinitely differentiable distribution.

The first derivative of the delta function is the distributional limit of the difference quotients:

$$\delta'(x) = \lim_{h\to 0} \frac{\delta(x+h) - \delta(x)}{h}$$

More properly, one has

$$\delta' = \lim_{h\to 0} \frac{1}{h}(\tau_h\delta - \delta)$$

where τ_h is the translation operator, defined on functions by $\tau_h\varphi(x) = \varphi(x + h)$, and on a distribution S by

$$(\tau_h S)[\varphi] = S[\tau_{-h}\varphi].$$

In the theory of electromagnetism, the first derivative of the delta function represents a point magnetic dipole situated at the origin. Accordingly, it is referred to as a dipole or the doublet function.

The derivative of the delta function satisfies a number of basic properties, including:

$$\frac{d}{dx}\delta(-x) = -\frac{d}{dx}\delta(x)$$

$$x\delta'(x) = -\delta(x).$$

Furthermore, the convolution of δ' with a compactly supported smooth function f is

$$\delta' * f = \delta * f' = f',$$

which follows from the properties of the distributional derivative of a convolution.

Higher Dimensions

More generally, on an open set U in the n-dimensional Euclidean space R^n, the Dirac delta distribution centred at a point $a - U$ is defined by

$$\delta_a[\phi] = \phi(a)$$

for all $\varphi - S(U)$, the space of all smooth compactly supported functions on U. If $\alpha = (\alpha_1,..., \alpha_n)$ is any multi-index and $-^{\alpha}$ denotes the associated mixed partial derivative operator, then the αth derivative $-^{\alpha}\delta_a$ of δ_a is given by

$$\langle \partial^{\alpha}\delta_a, \varphi \rangle = (-1)^{|\alpha|}\langle \delta_a, \partial^{\alpha}\varphi \rangle = (-1)^{|\alpha|}\partial^{\alpha}\varphi(x)\Big|_{x=a} \quad \text{for all } \varphi \in S(U).$$

That is, the αth derivative of δ_a is the distribution whose value on any test function φ is the αth derivative of φ at a (with the appropriate positive or negative sign).

The first partial derivatives of the delta function are thought of as double layers along the coordinate planes.

More generally, the normal derivative of a simple layer supported on a surface is a double layer supported on that surface, and represents a laminar magnetic monopole. Higher derivatives of the delta function are known in physics as multipoles.

Higher derivatives enter into mathematics naturally as the building blocks for the complete structure of distributions with point support. If S is any distribution on U supported on the set

$\{a\}$ consisting of a single point, then there is an integer m and coefficients c_α such that

$$S = \sum_{|\alpha| \le m} c_\alpha \partial^\alpha \delta_a .$$

Representations of the Delta Function

The delta function can be viewed as the limit of a sequence of functions

$$\delta(x) = \lim_{\varepsilon \to 0^+} \eta_\varepsilon(x),$$

where $\eta_\varepsilon(x)$ is sometimes called a nascent delta function. This limit is meant in a weak sense: either that

$$\lim_{\varepsilon \to 0^+} \int_{-\infty}^{\infty} \eta_\varepsilon(x) f(x) dx = f(0) \tag{5}$$

for all continuous functions f having compact support, or that this limit holds for all smooth functions f with compact support. The difference between these two slightly different modes of weak convergence is often subtle: the former is convergence in the vague topology of measures, and the latter is convergence in the sense of distributions.

Approximations to the Identity

Typically a nascent delta function η_ε can be constructed in the following manner. Let η be an absolutely integrable function on R of total integral 1, and define

$$\eta_\varepsilon(x) = \eta(x / \varepsilon) / \varepsilon.$$

In n dimensions, one uses instead the scaling

$$\eta_\varepsilon(x) = \eta(x / \varepsilon) / \varepsilon^n .$$

Then a simple change of variables shows that η_ε also has integral 1. One shows easily that (5) holds for all continuous compactly supported functions f, and so η_ε converges weakly to δ in the sense of measures.

The η_ε constructed in this way are known as an approximation to the identity. This terminology is because the space $L^1(\mathrm{R})$ of absolutely integrable functions is closed under the operation of convolution of functions: $f–g – L^1(\mathrm{R})$ whenever f and g are in $L^1(\mathrm{R})$. However, there is no identity in $L^1(\mathrm{R})$ for the convolution product: no element h such that $f–h = f$ for all

f. Nevertheless, the sequence η_ε does approximate such an identity in the sense that

$$f * \eta_\varepsilon \to f \quad \text{as } \varepsilon \to 0.$$

This limit holds in the sense of mean convergence (convergence in L^1). Further conditions on the η_ε, for instance that it be a mollifier associated to a compactly supported function, are needed to ensure pointwise convergence almost everywhere.

If the initial $\eta = \eta_1$ is itself smooth and compactly supported then the sequence is called a mollifier. The standard mollifier is obtained by choosing η to be a suitably normalised bump function, for instance

$$\eta(x) = \begin{cases} e^{-1/(1-|x|^2)} & \text{if } |x| < 1 \\ 0 & \text{if } |x| \geq 1. \end{cases}$$

In some situations such as numerical analysis, a piecewise linear approximation to the identity is desirable. This can be obtained by taking η_1 to be a hat function. With this choice of η_1, one has

$$\eta_\varepsilon(x) = \varepsilon^{-1} \max(1 - |x/\varepsilon|, 0)$$

which are all continuous and compactly supported, although not smooth and so not a mollifier.

Probabilistic Considerations

In the context of probability theory, it is natural to impose the additional condition that the initial η_1 in an approximation to the identity should be positive, as such a function then represents a probability distribution. Convolution with a probability distribution is sometimes favourable because it does not result in overshoot or undershoot, as the output is a convex combination of the input values, and thus falls between the maximum and minimum of the input function. Taking η_1 to be any probability distribution at all, and letting $\eta_\varepsilon(x) = \eta_1(x/\varepsilon)/\varepsilon$ as above will give rise to an approximation to the identity. In general this converges more rapidly to a delta function if, in addition, η has mean 0 and has small higher moments. For instance, if η_1 is the uniform distribution on $[-1/2, 1/2]$, also known as the rectangular function, then:

$$\eta_\varepsilon(x) = \frac{1}{\varepsilon}\operatorname{rect}\left(\frac{x}{\varepsilon}\right) = \begin{cases} \dfrac{1}{\varepsilon}, & -\dfrac{\varepsilon}{2} < x < \dfrac{\varepsilon}{2} \\ 0, & \text{otherwise.} \end{cases}$$

Another example is with the Wigner semicircle distribution

$$\eta_\varepsilon(x) = \begin{cases} \dfrac{2}{\pi\varepsilon^2}\sqrt{\varepsilon^2 - x^2}, & -\varepsilon < x < \varepsilon \\ 0, & \text{otherwise} \end{cases}$$

This is continuous and compactly supported, but not a mollifier because it is not smooth.

Semigroups

Nascent delta functions often arise as convolution semigroups. This amounts to the further constraint that the convolution of η_ε with η_δ must satisfy

$$\eta_\varepsilon * \eta_\delta = \eta_{\varepsilon+\delta}$$

for all $\varepsilon,\delta > 0$. Convolution semigroups in L^1 that form a nascent delta function are always an approximation to the identity in the above sense, however the semigroup condition is quite a strong restriction.

In practice, semigroups approximating the delta function arise as fundamental solutions or Green's functions to physically motivated elliptic or parabolic partial differential equations.

In the context of applied mathematics, semigroups arise as the output of a linear time-invariant system. Abstractly, if A is a linear operator acting on functions of x, then a convolution semigroup arises by solving the initial value problem

$$\begin{cases} \dfrac{\partial}{\partial t}\eta(t,x) = A\eta(t,x), & t > 0 \\ \lim\limits_{t\to 0^+}\eta(t,x) = \delta(x) \end{cases}$$

in which the limit is as usual understood in the weak sense. Setting $\eta_\varepsilon(x) = \eta(\varepsilon,x)$ gives the associated nascent delta function.

Some examples of physically important convolution semigroups arising from such a fundamental solution include the following.

The Heat Kernel:

The heat kernel, defined by

$$\eta_\varepsilon(x) = \frac{1}{\sqrt{2\pi\varepsilon}} e^{-x^2/2\varepsilon}$$

represents the temperature in an infinite wire at time $t > 0$, if a unit of heat energy is stored at the origin of the wire at time $t = 0$. This semigroup evolves according to the one-dimensional heat equation:

$$\frac{\partial u}{\partial t} = \frac{1}{2} \frac{\partial^2 u}{\partial x^2} . a$$

In probability theory, $\eta_\varepsilon(x)$ is a normal distribution of variance ε and mean 0.

It represents the probability density at time $t = \varepsilon$ of the position of a particle starting at the origin following a standard Brownian motion.

In this context, the semigroup condition is then an expression of the Markov property of Brownian motion.

In higher dimensional Euclidean space R^n, the heat kernel is

$$\eta_\varepsilon = \frac{1}{(2\pi\varepsilon)^{n/2}} e^{-x \cdot x/2\varepsilon},$$

and has the same physical interpretation, *mutatis mutandis*. It also represents a nascent delta function in the sense that $\eta_\varepsilon \to \delta$ in the distribution sense as $\varepsilon \to 0$.

The Poisson Kernel:

$$\eta_\varepsilon(x) = \frac{1}{\pi} \frac{\varepsilon}{\varepsilon^2 + x^2} = \int_{-\infty}^{\infty} e^{2\pi i \xi x - |\varepsilon \xi|} \, d\xi$$

is the fundamental solution of the Laplace equation in the upper half-plane. It represents the electrostatic potential in a semi-infinite plate whose potential along the edge is held at fixed at the delta function.

The Poisson kernel is also closely related to the Cauchy distribution. This semigroup evolves according to the equation

$$\frac{\partial u}{\partial t} = -(-\partial^2 / \partial x^2)^{1/2} u(t, x)$$

where the operator is rigorously defined as the Fourier multiplier

$$\mathcal{F}\left[(-\partial^2/\partial x^2)^{1/2} f\right](\xi) = |2\pi\xi|\, \mathcal{F}f(\xi).$$

Oscillatory Integrals

In areas of physics such as wave propagation and wave mechanics, the equations involved are hyperbolic and so may have more singular solutions.

As a result, the nascent delta functions that arise as fundamental solutions of the associated Cauchy problems are generally oscillatory integrals. An example, which comes from a solution of the Euler–Tricomi equation of transonic gas dynamics, is the rescaled Airy function

$$\mathrm{Ai}(x/\varepsilon^{1/3})/\varepsilon^{1/3}.$$

Although using the Fourier transform, it is easy to see that this generates a semigroup in some sense, it is not absolutely integrable and so cannot define a semigroup in the above strong sense. Many nascent delta functions constructed as oscillatory integrals only converge in the sense of distributions (an example is the Dirichlet kernel below), rather than in the sense of measures.

Another example is the Cauchy problem for the wave equation in R^{1+1}:

$$c^{-2}\frac{\partial^2 u}{\partial t^2} - \Delta u = 0$$

$$u = 0, \qquad \frac{\partial u}{\partial t} = \delta \quad \text{when } t=0.$$

The solution u represents the displacement from equilibrium of an infinite elastic string, with an initial disturbance at the origin.

Other approximations to the identity of this kind include the sinc function

$$\eta_\varepsilon(x) = \frac{1}{\pi x}\sin\left(\frac{x}{\varepsilon}\right) = \frac{1}{2\pi}\int_{-1/\varepsilon}^{1/\varepsilon}\cos(kx)\,dk$$

and the Bessel function

$$\eta_\varepsilon(x) = \frac{1}{\varepsilon}J_{1/\varepsilon}\left(\frac{x+1}{\varepsilon}\right).$$

Plane Wave Decomposition

One approach to the study of a linear partial differential equation

$$L[u] = f,$$

where L is a differential operator on R^n, is to seek first a fundamental solution, which is a solution of the equation

$$L[u] = \delta.$$

When L is particularly simple, this problem can often be resolved using the Fourier transform directly (as in the case of the Poisson kernel and heat kernel already mentioned). For more complicated operators, it is sometimes easier first to consider an equation of the form

$$L[u] = h$$

where h is a plane wave function, meaning that it has the form

$$h = h(x \cdot \xi)$$

for some vector ∂. Such an equation can be resolved (if the coefficients of L are analytic functions) by the Cauchy–Kovalevskaya theorem or (if the coefficients of L are constant) by quadrature.

So, if the delta function can be decomposed into plane waves, then one can in principle solve linear partial differential equations.

Such a decomposition of the delta function into plane waves was part of a general technique first introduced essentially by Johann Radon, and then developed in this form by Fritz John (1955). Choose k so that $n + k$ is an even integer, and for a real number s, put

$$g(s) = \operatorname{Re}\left[\frac{-s^k \log(-is)}{k!(2\pi i)^n}\right] = \begin{cases} \dfrac{|s|^k}{4k!(2\pi i)^{n-1}} & n \text{ odd} \\ -\dfrac{|s|^k \log|s|}{k!(2\pi i)^n} & n \text{ even.} \end{cases}$$

Then δ is obtained by applying a power of the Laplacian to the integral with respect to the unit sphere measure $d\omega$ of $g(x \cdot \partial)$ for ∂ in the unit sphere S^{n-1}:

$$\delta(x) = \Delta_x^{(n+k)/2} \int_{S^{n-1}} g(x \cdot \xi) d\omega_\xi .$$

The Laplacian here is interpreted as a weak derivative, so that this equation is taken to mean that, for any test function φ,

$$\varphi(x) = \int_{\mathbf{R}^n} \varphi(y) dy \Delta_x^{(n+k)/2} \int_{S^{n-1}} g((x-y) \cdot \xi) d\omega_\xi .$$

The result follows from the formula for the Newtonian potential (the fundamental solution of Poisson's equation). This is essentially a form of the inversion formula for the Radon transform, because it recovers the value of $\varphi(x)$ from its integrals over hyperplanes. For instance, if n is odd and $k = 1$, then the integral on the right hand side is

$$c_n \Delta_x^{(n+1)/2} \int \int_{S^{n-1}} \varphi(y) \,|\, (y-x) \cdot \xi \,| d\omega_\xi \, dy$$

$$= c_n \Delta_x^{(n+1)/2} \int_{S^{n-1}} d\omega_\xi \int_{-\infty}^{\infty} |p| \, R\varphi(\xi, p + x \cdot \xi) dp$$

where $R\varphi(\partial,p)$ is the Radon transform of φ:

$$R\varphi(\xi, p) = \int_{x \cdot \xi = p} f(x) d^{n-1} x.$$

An alternative equivalent expression of the plane wave decomposition, from Gel'fand & Shilov (1966–1968, I, §3.10), is

$$\delta(x) = \frac{(n-1)!}{(2\pi i)^n} \int_{S^{n-1}} (x \cdot \xi)^{-n} d\omega_\xi$$

for n even, and

$$\delta(x) = \frac{1}{2(2\pi i)^{n-1}} \int_{S^{n-1}} \delta^{(n-1)} (x \cdot \xi) d\omega_\xi$$

for n odd.

Fourier Kernels

In the study of Fourier series, a major question consists of determining whether and in what sense the Fourier series associated with a periodic function converges to the function. The nth partial sum of the Fourier series of a function f of period 2π is defined by convolution (on the interval $[-\pi, \pi]$) with the Dirichlet kernel:

$$D_N(x) = \sum_{n=-N}^{N} e^{inx} = \frac{\sin\left((N + \frac{1}{2})x\right)}{\sin(x/2)}.$$

Thus,

$$s_N(f)(x) = D_N * f(x) = \sum_{n=-N}^{N} a_n e^{inx}$$

where

$$a_n = \frac{1}{2\pi}\int_{-\pi}^{\pi} f(y)e^{-iny}\,dy.$$

A fundamental result of elementary Fourier series states that the Dirichlet kernel tends to the a multiple of the delta function as $N \to \infty$ This is interpreted in the distribution sense, that

$$s_N(f)(0) = \int_{\mathbf{R}} D_N(x)f(x)dx \to 2\pi f(0)$$

for every compactly supported *smooth* function f. Thus, formally one has

$$\delta(x) = \frac{1}{2\pi}\sum_{n=-\infty}^{\infty} e^{inx}$$

on the interval $[-\pi,\pi]$.

In spite of this, the result does not hold for all compactly supported *continuous* functions: that is D_N does not converge weakly in the sense of measures.

The lack of convergence of the Fourier series has led to the introduction of a variety of summability methods in order to produce convergence.

The method of Cesàro summation leads to the Fejér kernel

$$F_N(x) = \sum_{n=0}^{N} D_n(x) = \frac{1}{N}\left(\frac{\sin\frac{Nx}{2}}{\sin\frac{x}{2}}\right)^2.$$

The Fejér kernels tend to the delta function in a stronger sense that

$$\int_{\mathbf{R}} F_N(x)f(x)dx \to 2\pi f(0)$$

for every compactly supported *continuous* function f. The implication is that the Fourier series of any continuous function is Cesàro summable to the value of the function at every point.

Hilbert Space Theory

The Dirac delta distribution is a densely defined unbounded linear functional on the Hilbert space L^2 of square integrable functions. Indeed, smooth compactly support functions are dense in L^2, and the action of the delta distribution on such functions is well-defined. In many applications, it is possible to identify subspaces of L^2 and to give a stronger topology on which the delta function defines a bounded linear functional.

Sobolev Spaces

The Sobolev embedding theorem for Sobolev spaces on the real line R implies that any square-integrable function f such that

$$\|f\|_{H^1}^2 = \int_{-\infty}^{\infty} \left|\hat{f}(\xi)\right|^2 \left(1+|\xi|^2\right) d\xi < \infty$$

is automatically continuous, and satisfies in particular

$$\delta[f] = |f(0)| < C\|f\|_{H^1}.$$

Thus δ is a bounded linear functional on the Sobolev space H^1. Equivalently δ is an element of the continuous dual space H^{-1} of H^1. More generally, in n dimensions, one has $\delta - H^s(\mathrm{R}^n)$ provided $s > n/2$.

Spaces of Holomorphic Functions

In complex analysis, the delta function enters via Cauchy's integral formula which asserts that if D is a domain in the complex plane with smooth boundary, then

$$f(z) = \frac{1}{2\pi i} \oint_{\partial D} \frac{f(\zeta)d\zeta}{\zeta - z}, \quad z \in D$$

for all holomorphic functions f in D that are continuous on the closure of D. As a result, the delta function δ_z is represented on this class of holomorphic functions by the Cauchy integral:

$$\delta_z[f] = f(z) = \frac{1}{2\pi i} \oint_{\partial D} \frac{f(\zeta)d\zeta}{\zeta - z}.$$

More generally, let $H^2(-D)$ be the Hardy space consisting of the closure in $L^2(-D)$ of all holomorphic functions in D continuous up to the boundary of D. Then functions in $H^2(-D)$ uniquely extend to holomorphic functions in D, and the Cauchy integral formula continues to hold. In particular for $z - D$, the delta function δ_z is a continuous linear functional on $H^2(-D)$.

This is a special case of the situation in several complex variables in which, for smooth domains D, the Szegõ kernel plays the role of the Cauchy integral.

Resolutions of the Identity

Given a complete orthonormal basis set of functions $\{\varphi_n\}$ in a separable Hilbert space, for example, the normalised eigenvectors of a compact self-adjoint operator, any vector f can be expressed as:

$$f = \sum_{n=1}^{\infty} \alpha_n \varphi_n .$$

The coefficients $\{\alpha_n\}$ are found as:

$$\alpha_n = \langle \varphi_n, f \rangle ,$$

which may be represented by the notation:

$$\alpha_n = \varphi_n^{\dagger} f ,$$

a form of the bra-ket notation of Dirac. Adopting this notation, the expansion of f takes the dyadic form:

$$f = \sum_{n=1}^{\infty} \varphi_n \left(\varphi_n^{\dagger} f \right).$$

Letting I denote the identity operator on the Hilbert space, the expression

$$I = \sum_{n=1}^{\infty} \varphi_n \varphi_n^{\dagger},$$

is called a resolution of the identity. When the Hilbert space is the space $L^2(D)$ of square-integrable functions on a domain D, the quantity:

$$\varphi_n \varphi_n^{\dagger},$$

is an integral operator, and the expression for f can be rewritten as:

$$f(x) = \sum_{n=1}^{\infty} \int_D \left(\varphi_n(x) \varphi_n^{*}(\xi) \right) f(\xi) d\xi.$$

The right-hand side converges to f in the L^2 sense. It need not hold in a pointwise sense, even when f is a continuous function.

Nevertheless, it is common to abuse notation and write

$$f(x) = \int \delta(x - \xi) f(\xi) d\xi,$$

resulting in the representation of the delta function:

$$\delta(x - \xi) = \sum_{n=1}^{\infty} \varphi_n(x)\varphi_n^*(\xi).$$

With a suitable rigged Hilbert space $(\Phi, L^2(D), \Phi^-)$ where $\Phi \subset L^2(D)$ contains all compactly supported smooth functions, this summation may converge in Φ^*, depending on the properties of the basis φ_n.

In most cases of practical interest, the orthonormal basis comes from an integral or differential operator, in which case the series converges in the distribution sense.

Infinitesimal Delta Functions

Cauchy used an infinitesimal α to write down a unit impulse, infinitely tall and narrow Dirac-type delta function δ_α satisfying $\int F(x)\delta_\alpha(x) = F(0)$ in a number of articles in 1827. Cauchy defined an infinitesimal in Cours d'Analyse (1827) in terms of a sequence tending to zero.

Namely, such a null sequence becomes an infinitesimal in Cauchy's and Lazare Carnot's terminology.

Modern set-theoretic approaches allow one to define infinitesimals via the ultrapower construction, where a null sequence becomes an infinitesimal in the sense of an equivalence class modulo a relation defined in terms of a suitable ultrafilter.

The article by Yamashita (2007) contains a bibliography on modern Dirac delta functions in the context of an infinitesimal-enriched continuum provided by the hyperreals.

Here the Dirac delta can be given by an actual function, having the property that for every real function F one has $\int F(x)\delta_\alpha(x) = F(0)$ as anticipated by Fourier and Cauchy.

Dirac Comb

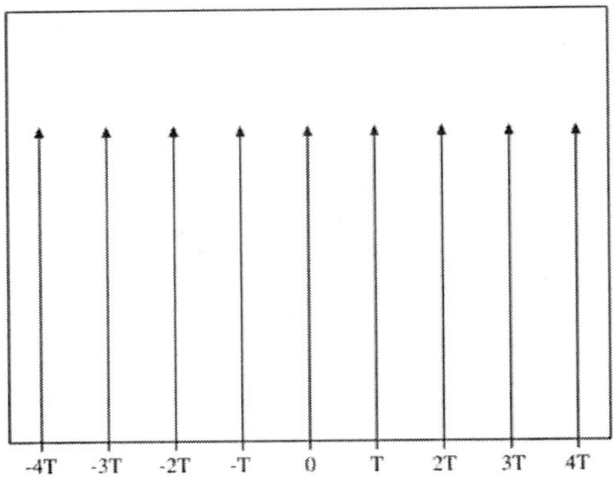

Figure: *A Dirac comb is an infinite series of Dirac delta functions spaced at intervals of* T

A so-called uniform "pulse train– of Dirac delta measures, which is known as a Dirac comb, or as the Shah distribution, creates a sampling function, often used in digital signal processing (DSP) and discrete time signal analysis. The Dirac comb is given as the infinite sum, whose limit is understood in the distribution sense,

$$\Delta(x) = \sum_{n=-\infty}^{\infty} \delta(x-n),$$

which is a sequence of point masses at each of the integers.

Up to an overall normalising constant, the Dirac comb is equal to its own Fourier transform. This is significant because if f is any Schwartz function, then the periodisation of f is given by the convolution

$$(f * \Delta)(x) = \sum_{n=-\infty}^{\infty} f(x-n).$$

In particular,

$$(f * \Delta)^{\wedge} = \hat{f}\hat{\Delta} = \hat{f}\Delta$$

is precisely the Poisson summation formula.

Sokhatsky–Weierstrass Theorem

The Sokhatsky–Weierstrass theorem, important in quantum mechanics, relates the delta function to the distribution p.v.$1/x$, the Cauchy principal value of the function $1/x$, defined by

$$\left\langle \text{p.v.}\frac{1}{x}, \phi \right\rangle = \lim_{\epsilon\to 0^+}\int_{|x|>\epsilon}\frac{\phi(x)}{x}dx.$$

Sokhatsky's formula states that

$$\lim_{\epsilon\to 0^+}\frac{1}{x \pm i\epsilon} = \text{p.v.}\frac{1}{x} \mp i\pi\delta(x),$$

Here the limit is understood in the distribution sense, that for all compactly supported smooth functions f,

$$\lim_{\epsilon\to 0^+}\int_{-\infty}^{\infty}\frac{f(x)}{x \pm i\epsilon}dx = \mp i\pi f(0) + \lim_{\epsilon\to 0^+}\int_{|x|>\epsilon}\frac{f(x)}{x}dx.$$

Relationship to the Kronecker Delta

The Kronecker delta δ_{ij} is the quantity defined by

$$\delta_{ij} = \begin{cases} 1 & i = j \\ 0 & i \neq j \end{cases}$$

for all integers i, j. This function then satisfies the following analog of the sifting property: if $(a_i)_{i\in\mathbb{Z}}$ be any doubly infinite sequence, then

$$\sum_{i=-\infty}^{\infty} a_i \delta_{ik} = a_k.$$

Similarly, for any real or complex valued continuous function f on $\mathbb{R}$, the Dirac delta satisfies the sifting property

$$\int_{-\infty}^{\infty} f(x)\delta(x - x_0)dx = f(x_0).$$

This exhibits the Kronecker delta function as a discrete analog of the Dirac delta function.

Application to Quantum Mechanics

We give an example of how the delta function is expedient in quantum mechanics. The wave function of a particle gives the probability amplitude of finding a particle within a given region of space. Wave functions are assumed to be elements of the Hilbert space L^2 of square-integrable functions, and the total probability of finding a particle within a given interval is the integral of the magnitude of the wave function squared over the interval. A set $\{\varphi_n\}$ of wave functions is orthonormal if they are normalised by

$$\langle \phi_n | \phi_m \rangle = \delta_{nm}$$

where δ here refers to the Kronecker delta. A set of orthonormal wave functions is complete in the space of square-integrable functions if any wave function ψ can be expressed as a combination of the φ_n:

$$\psi = \sum c_n \phi_n,$$

with $c_n = \langle \phi_n | \psi \rangle$. Complete orthonormal systems of wave functions appear naturally as the eigenfunctions of the Hamiltonian (of a bound system) in quantum mechanics that measures the energy levels, which are called the eigenvalues. The set of eigenvalues, in this case, is known as the spectrum of the Hamiltonian. In bra-ket notation, as above, this equality implies the resolution of the identity:

$$I = \sum |\phi_n\rangle\langle\phi_n|.$$

Here the eigenvalues are assumed to be discrete, but the set of eigenvalues of an observable may be continuous rather than discrete. An example is the position observable, $Q\psi(x) = x\psi(x)$. The spectrum of the position (in one dimension) is the entire real line, and is called a continuous spectrum. However, unlike the Hamiltonian, the position operator lacks proper eigenfunctions. The conventional way to overcome this shortcoming is to widen the class of available functions by allowing distributions as well: that is, to replace the Hilbert space of quantum mechanics by an appropriate rigged Hilbert space. In this context, the position operator has a complete set of eigen-distributions, labelled by the points y of the real line, given by

$$\phi_y(x) = \delta(x - y).$$

The eigenfunctions of position are denoted by $\phi_y = |y\rangle$ in Dirac notation, and are known as position eigenstates.

Similar considerations apply to the eigenstates of the momentum operator, or indeed any other self-adjoint unbounded operator P on the Hilbert space, provided the spectrum of P

is continuous and there are no degenerate eigenvalues. In that case, there is a set Ω of real numbers (the spectrum), and a collection φ_y of distributions indexed by the elements of Ω, such that

$$P\phi_y = y\phi_y.$$

That is, φ_y are the eigenvectors of P. If the eigenvectors are normalised so that

$$\langle \phi_y, \phi_{y'} \rangle = \delta(y - y')$$

in the distribution sense, then for any test function ψ,

$$\psi(x) = \int_\Omega c(y)\phi_y(x)dy$$

where

$$c(y) = \langle \psi, \phi_y \rangle.$$

That is, as in the discrete case, there is a resolution of the identity

$$I = \int_\Omega |\phi_y\rangle\langle\phi_y|dy$$

where the operator-valued integral is again understood in the weak sense. If the spectrum of P has both continuous and discrete parts, then the resolution of the identity involves a summation over the discrete spectrum *and* an integral over the continuous spectrum.

The delta function also has many more specialised applications in quantum mechanics, such as the delta potential models for a single and double potential well.

Solution Methods

Separation of Variables

In mathematics, separation of variables is any of several methods for solving ordinary and partial differential equations, in which algebra allows one to rewrite an equation so that each of two variables occurs on a different side of the equation.

Ordinary Differential Equations (ODE)

Suppose a differential equation can be written in the form

$$\frac{d}{dx} f(x) = g(x)h(f(x)), \tag{1}$$

which we can write more simply by letting $y = f(x)$:

$$\frac{dy}{dx} = g(x)h(y).$$

As long as $h(y) \neq 0$, we can rearrange terms to obtain:

$$\frac{dy}{h(y)} = g(x)dx,$$

so that the two variables x and y have been separated. dx (and dy) can be viewed, at a simple level, as just a convenient notation, which provides a handy mnemonic aid for assisting with manipulations.

A formal definition of dx as a differential (infinitesimal) is somewhat advanced.

Alternative Notation

Some who dislike Leibniz's notation may prefer to write this as

$$\frac{1}{h(y)}\frac{dy}{dx} = g(x),$$

but that fails to make it quite as obvious why this is called "separation of variables".

Integrating both sides of the equation with respect to x, we have

$$\int \frac{1}{h(y)}\frac{dy}{dx}dx = \int g(x)dx, \tag{2}$$

or equivalently,

$$\int \frac{1}{h(y)}dy = \int g(x)dx$$

because of the substitution rule for integrals.

If one can evaluate the two integrals, one can find a solution to the differential equation. Observe that this process effectively allows us to treat the derivative $\frac{dy}{dx}$ as a fraction which can be separated. This allows us to solve separable differential equations more conveniently, as demonstrated in the example below.

(Note that we do not need to use two constants of integration, in equation (2) as in

$$\int \frac{1}{h(y)} dy + C_1 = \int g(x) dx + C_2,$$

because a single constant $C = C_2 - C_1$ is equivalent.)

Partial Differential Equations

The method of separation of variables is also used to solve a wide range of linear partial differential equations with boundary and initial conditions, such as heat equation, wave equation, Laplace equation and Helmholtz equation.

Homogeneous Case

Consider the one-dimensional heat equation. The equation is

$$\frac{\partial u}{\partial t} - \alpha \frac{\partial^2 u}{\partial x^2} = 0 \tag{1}$$

The boundary condition is homogeneous, that is

$$u\big|_{x=0} = u\big|_{x=L} = 0 \tag{2}$$

Let us attempt to find a solution which is not identically zero satisfying the boundary conditions but with the following property: u is a product in which the dependence of u on x, t is separated, that is:

$$u(x,t) = X(x)T(t). \tag{3}$$

Substituting u back into equation,

$$\frac{T'(t)}{\alpha T(t)} = \frac{X''(x)}{X(x)}. \tag{4}$$

Since the right hand side depends only on x and the left hand side only on t, both sides are equal to some constant value $-\lambda$. Thus:

$$T'(t) = -\lambda \alpha T(t), \tag{5}$$

and

$$X''(x) = -\lambda X(x). \tag{6}$$

$-\lambda$ here is the eigenvalue for both differential operators, and *T(t)* and *X(x)* are corresponding eigenfunctions.

We will now show that solutions for *X(x)* for values of λ d– 0 cannot occur:

Suppose that $\lambda < 0$. Then there exist real numbers *B*, *C* such that

$$X(x) = Be^{\sqrt{-\lambda}\,x} + Ce^{-\sqrt{-\lambda}\,x}.$$

From (2) we get

$$X(0) = 0 = X(L), \tag{7}$$

and therefore $B = 0 = C$ which implies u is identically 0.

Suppose that $\lambda = 0$. Then there exist real numbers *B*, *C* such that

$$X(x) = Bx + C.$$

From (7) we conclude in the same manner as in 1 that u is identically 0.

Therefore, it must be the case that $\lambda > 0$. Then there exist real numbers *A*, *B*, *C* such that

$$T(t) = Ae^{-\lambda\alpha t},$$

and

$$X(x) = B\sin(\sqrt{\lambda}\,x) + C\cos(\sqrt{\lambda}\,x).$$

From (7) we get $C = 0$ and that for some positive integer n,

$$\sqrt{\lambda} = n\frac{\pi}{L}.$$

This solves the heat equation in the special case that the dependence of u has the special form of (3).

In general, the sum of solutions to (1) which satisfy the boundary conditions (2) also satisfies (1) and (3). Hence a complete solution can be given as

$$u(x,t) = \sum_{n=1}^{\infty} D_n \sin\frac{n\pi x}{L}\exp\left(-\frac{n^2\pi^2\alpha t}{L^2}\right),$$

where D_n are coefficients determined by initial condition.

Given the initial condition

$$u\big|_{t=0} = f(x),$$

we can get

$$f(x) = \sum_{n=1}^{\infty} D_n \sin\frac{n\pi x}{L}.$$

This is the sine series expansion of *f(x)*. Multiplying both sides with $\sin\frac{n\pi x}{L}$ and integrating over *[0,L]* result in

$$D_n = \frac{2}{L}\int_0^L f(x)\sin\frac{n\pi x}{L}dx.$$

This method requires that the eigenfunctions of *x*, here $\left\{\sin\frac{n\pi x}{L}\right\}_{n=1}^{\infty}$, are orthogonal and complete. In general this is guaranteed by Sturm-Liouville theory.

Matrices

The matrix form of the separation of variables is the Kronecker sum.

As an example we consider the 2D discrete Laplacian on a regular grid:

$$L = D_{xx} \oplus D_{yy} = D_{xx} \otimes I + I \otimes D_{yy},$$

where D_{xx} and D_{yy} are 1D discrete Laplacians in the *x*- and *y*-directions, correspondingly, and I are the identities of appropriate sizes.

Integrating Factor

In mathematics, an integrating factor is a function that is chosen to facilitate the solving of a given equation involving differentials.

It is commonly used to solve ordinary differential equations, but is also used within multivariable calculus when multiplying through by an integrating factor allows an inexact differential to be made into an exact differential (which can then be integrated to give a scalar field). This is especially useful in thermodynamics where temperature becomes the integrating factor that makes entropy an exact differential.

Use in Solving First Order Linear Ordinary Differential Equations

Consider an ordinary differential equation of the form

$$y' + P(x)y = Q(x) \tag{1}$$

Consider a function $M(x)$. We multiply both sides of (1) by $M(x)$:

$$M(x)y' + M(x)P(x)y = M(x)Q(x). \tag{2}$$

We want the left hand side to be in the form of the derivative of a product, such that (2) can be written as

$$(M(x)y)' = M(x)Q(x). \tag{3}$$

The left hand side in (3) can now be integrated

$$yM(x) = \int Q(x)M(x)dx,$$

We can now solve for y,

$$y = \frac{\int Q(x)M(x)dx}{M(x)}.$$

Applying the product rule to the left hand side of (3) and equating to the left hand side of (2)

$$M'(x)y + M(x)y' = M(x)y' + M(x)P(x)y.$$

From which it is clear that $M(x)$ obeys the differential equation :

$$M'(x) = M(x)P(x). \tag{4}$$

$$\frac{M'(x)}{M(x)} = P(x). \tag{5}$$

Keeping in mind that for:

$$F(x) = e^{G(x)}.$$

$$F'(x) = G'(x) * e^{G(x)}.$$

$$F'(x) = G'(x) * F(x).$$

We can see that when we let:

$$P(x) = G'(x)$$

We can solve (5):

$$\int P(x)dx = G(x)$$

$$e^{\int P(x)dx} = e^{G(x)}$$

$$e^{\int P(x)dx} = M(x)$$

$M(x)$ is called an integrating factor.

General Use

The term "integrating factor" is synonymous with the solution of first order linear equations.

One should bear in mind, however, that an integrating factor is any expression that a differential equation is multiplied by to facilitate integration, and that it is by no means restricted to first order linear equations.

For example, the nonlinear second order equation

$$\frac{d^2y}{dt^2} = Ay^{2/3}$$

admits $\frac{dy}{dt}$ as an integrating factor:

$$\frac{d^2y}{dt^2}\frac{dy}{dt} = Ay^{2/3}\frac{dy}{dt}.$$

To integrate, note that both sides of the equation may be expressed as derivatives by going backwards with the chain rule:

$$\frac{d}{dt}\left(\frac{1}{2}\left(\frac{dy}{dt}\right)^2\right) = \frac{d}{dt}\left(A\frac{3}{5}y^{5/3}\right).$$

Therefore

$$\left(\frac{dy}{dt}\right)^2 = \frac{6A}{5}y^{5/3} + C_0.$$

This form may be more useful, depending on application. Performing a separation of variables will give:

$$\int \frac{dy}{\sqrt{\frac{6A}{5} y^{5/3} + C_0}} = t + C_1;$$

this is an implicit solution which involves a nonelementary integral.

Though likely too obscure to be useful, this is a general solution.

Also, because the previous equation is first order, it could be used for numeric solution in favour of the original equation.

Chapter 3

Forcing Function

In a system of differential equations used to describe a time-dependent process, a forcing function is a function that appears in the equations and is only a function of time, not of any of the other variables. In effect, it is a constant for each value of t.

In the more general case, any nonhomogeneous source function in any variable can be described as a forcing function, and the resulting solution can often be determined using a superposition of linear combinations of the homogeneous solutions and the forcing term.

Laplace Transform Applied to Differential Equations

The Laplace transform is a powerful integral transform used to switch a function from the time domain to the s-domain. The Laplace transform can be used in some cases to solve linear differential equations with given initial conditions.

First consider the following property of the Laplace transform:

$$\mathcal{L}\{f'\} = s\mathcal{L}\{f\} - f(0)$$

$$\mathcal{L}\{f''\} = s^2\mathcal{L}\{f\} - sf(0) - f'(0)$$

One by induction can prove that

$$\mathcal{L}\{f^{(n)}\} = s^n\mathcal{L}\{f\} - \sum_{i=1}^{n} s^{n-i} f^{(i-1)}(0)$$

Now we consider the following differential equation:

$$\sum_{i=0}^{n} a_i f^{(i)}(t) = \phi(t)$$

with given initial conditions

$$f^{(i)}(0) = c_i$$

Using the linearity of the Laplace transform it is equivalent to rewrite the equation as

$$\sum_{i=0}^{n} a_i \mathcal{L}\{f^{(i)}(t)\} = \mathcal{L}\{\phi(t)\}$$

obtaining

$$\mathcal{L}\{f(t)\}\sum_{i=0}^{n} a_i s^i - \sum_{i=1}^{n}\sum_{j=1}^{i} a_i s^{i-j} f^{(j-1)}(0) = \mathcal{L}\{\phi(t)\}$$

Solving the equation for $\mathcal{L}\{f(t)\}$ and substituting $f^{(i)}(0)$ with c_i one obtains

$$\mathcal{L}\{f(t)\} = \frac{\mathcal{L}\{\phi(t)\} + \sum_{i=1}^{n}\sum_{j=1}^{i} a_i s^{i-j} c_{j-1}}{\sum_{i=0}^{n} a_i s^i}$$

The solution for $f(t)$ is obtained by applying the inverse Laplace transform to $\mathcal{L}\{f(t)\}$.

Note that if the initial conditions are all zero, i.e.

$$f^{(i)}(0) = c_i = 0 \quad \forall i \in \{0,1,2,\dots n\}$$

then the formula simplifies to

$$f(t) = \mathcal{L}^{-1}\left\{\frac{\mathcal{L}\{\phi(t)\}}{\sum_{i=0}^{n} a_i s^i}\right\}$$

An Example

We want to solve

$$f''(t) + 4f(t) = \sin(2t)$$

with initial conditions $f(0) = 0$ and $f2$ (0)=0.

We note that

$$\phi(t) = \sin(2t)$$

and we get

$$\mathcal{L}\{\phi(t)\} = \frac{2}{s^2 + 4}$$

The equation is then equivalent to

$$s^2\mathcal{L}\{f(t)\} - sf(0) - f'(0) + 4\mathcal{L}\{f(t)\} = \mathcal{L}\{\phi(t)\}$$

We deduce

$$\mathcal{L}\{f(t)\} = \frac{2}{(s^2 + 4)^2}$$

Now we apply the Laplace inverse transform to get

$$f(t) = \frac{1}{8}\sin(2t) - \frac{t}{4}\cos(2t)$$

Liouville's Theorem (Differential Algebra)

In mathematics, Liouville's theorem, originally formulated by Joseph Liouville in the 1830s and 1840s, places an important restriction on antiderivatives that can be expressed as elementary functions.

The antiderivatives of certain elementary functions cannot themselves be expressed as elementary functions. A standard example of such a function is e^{-x^2}, whose antiderivative is (with a multiplier of a constant) the error function, familiar from statistics. Other examples include the functions $\frac{\sin(x)}{x}$ and x^x.

Liouville's theorem states that elementary antiderivatives, if they exist, must be in the same differential field as the function, plus possibly a finite number of logarithms.

Definitions

For any differential field F, there is a subfield

$$\mathrm{Con}(F) = \{f \text{ in } F \mid Df = 0\},$$

called the constants of F.

Given two differential fields F and G, G is called a logarithmic extension of F if G is a simple transcendental extension of F (i.e. $G = F(t)$ for some transcendental t) such that

$$Dt = Ds/s \text{ for some } s \text{ in } F.$$

This has the form of a logarithmic derivative. Intuitively, one may think of t as the logarithm of some element s of F, in which case, this condition is analogous to the ordinary chain rule.

But it must be remembered that F is not necessarily equipped with a unique logarithm; one might adjoin many "logarithm-like" extensions to F. Similarly, an exponential extension is a simple transcendental extension that satisfies

$$Dt = t\,Ds.$$

With the above caveat in mind, this element may be thought of as an exponential of an element s of F. Finally, G is called an elementary differential extension of F if there is a finite chain of subfields from F to G where each extension in the chain is either algebraic, logarithmic, or exponential.

Basic Theorem

Suppose F and G are differential fields, with Con(F) = Con(G), and that G is an elementary differential extension of F. Let a be in F, y in G, and suppose $Dy = a$ (in words, suppose that G contains an antiderivative of a). Then there exist $c_1,\ldots, c_n$ in Con(F), $u_1, \ldots, u_n$, v in F such that

$$a = c_1 \frac{Du_1}{u_1} + \cdots + c_n \frac{Du_n}{u_n} + Dv.$$

In other words, the only functions that have "elementary antiderivatives" (i.e. antiderivatives living in, at worst, an elementary differential extension of F) are those with this form prescribed by the theorem. Thus, on an intuitive level, the theorem states that the only elementary antiderivatives are the "simple" functions plus a finite number of logarithms of "simple" functions.

A proof of Liouville's theorem can be found in section 12.4 of Geddes, et al.

Examples

As an example, the field C(x) of rational functions in a single variable has a derivation given by the standard derivative with respect to that variable. The constants of this field are just the complex numbers C.

The function $\frac{1}{x}$, which exists in C(x), does not have an antiderivative in C(x). Its antiderivatives ln x + C do, however, exist in the logarithmic extension C(x, ln x).

Likewise, the function $\frac{1}{x^2+1}$ does not have an antiderivative in C(x). Its antiderivatives $\tan^{-1}(x) + C$ do not seem to satisfy the requirements of the theorem, since they are not (apparently) sums of rational functions and logarithms of rational functions.

However, a calculation with Euler's formula shows that in fact the antiderivatives can be written in the required manner (as logarithms of rational functions).

$$\begin{aligned}
e^{i\theta} &= \cos\theta + i\sin\theta \\
e^{-i\theta} &= \cos\theta - i\sin\theta \\
e^{2i\theta} &= \frac{e^{i\theta}}{e^{-i\theta}} = \frac{\cos\theta + i\sin\theta}{\cos\theta - i\sin\theta} \\
&= \frac{1 + i\tan\theta}{1 - i\tan\theta} \\
\text{[8}pt\text{]}2i\theta &= \ln\frac{1 + i\tan\theta}{1 - i\tan\theta} \\
\text{[8}pt\text{]}2i\tan^{-1}x &= \ln\frac{1 + ix}{1 - ix} \\
\text{[8}pt\text{]}\tan^{-1}x &= -\frac{1}{2}i\ln\frac{1 + ix}{1 - ix}
\end{aligned}$$

Relationship with Differential Galois Theory

Liouville's theorem is sometimes presented as a theorem in differential Galois theory, but this is not strictly true.

The theorem can be proved without any use of Galois theory. Furthermore, the Galois group of a simple antiderivative is either trivial (if no field extension is required to express it), or is simply the additive group of the constants (corresponding to the constant of integration).

Thus, an antiderivative's differential Galois group does not encode enough information to determine if it can be expressed using elementary functions, the major condition of Liouville's theorem.

Variation of Parameters

In mathematics, variation of parameters, also known as variation of constants, is a general method to solve inhomogeneous linear ordinary differential equations. It was developed by Joseph Louis Lagrange.

For first-order inhomogeneous linear differential equations it is usually possible to find solutions via integrating factors or undetermined coefficients with considerably less effort, although those methods leverage heuristics that involve guessing and don't work for all inhomogenous linear differential equations.

Variation of parameters extends to linear partial differential equations as well, specifically to inhomogeneous problems for linear evolution equations like the heat equation, wave equation, and vibrating plate equation.

In this setting, the method is more often known as Duhamel's principle, named after Jean-Marie Duhamel who first applied the method to solve the inhomogeneous heat equation. Sometimes variation of parameters itself is called Duhamel's principle and vice-versa.

Description of Method

Given an ordinary non-homogeneous linear differential equation of order n

$$y^{(n)}(x)+\sum_{i=0}^{n-1} a_i(x)y^{(i)}(x)=b(x). \tag{i}$$

let $y_1(x),\ldots,y_n(x)$ be a fundamental system of the corresponding homogeneous equation

$$y^{(n)}(x)+\sum_{i=0}^{n-1} a_i(x)y^{(i)}(x)=0. \tag{ii}$$

Then a particular solution to the non-homogeneous equation is given by

$$y_p(x)=\sum_{i=1}^{n} c_i(x)y_i(x) \tag{iii}$$

where the $c_i(x)$ are continuous functions which satisfy the equations

$$\sum_{i=1}^{n} c_{i'}(x) y_i^{(j)}(x) = 0, \quad j = 0, \ldots, n-2. \tag{iv}$$

By substituting (iii) into (i) and applying (iv) it follows that

$$\sum_{i=1}^{n} c_{i'}(x) y_i^{(n-1)}(x) = b(x). \tag{v}$$

The linear system (v) of n equations can then be solved using Cramer's rule yielding

$$c_{i'}(x) = \frac{W_i(x)}{W(x)}, \quad i = 1, \ldots, n$$

where $W(x)$ is the Wronskian determinant of the fundamental system and $W_i(x)$ is the Wronskian determinant of the fundamental system with the i-th column replaced by $(0, 0, \ldots, b(x))$.

The particular solution to the non-homogeneous equation can then be written as

$$\sum_{i=1}^{n} \int \frac{W_i(x)}{W(x)} dx \, y_i(x).$$

Examples:

Specific second order equation

Let us solve

$$y'' + 4y' + 4y = \cosh x.$$

We want to find the general solution to the differential equation, that is, we want to find solutions to the homogeneous differential equation

$$y'' + 4y' + 4y = 0.$$

From the characteristic equation

$$\lambda^2 + 4\lambda + 4 = (\lambda + 2)^2 = 0$$

$$\lambda = -2.$$

Since we have a repeated root, we have to introduce a factor of x for one solution to ensure linear independence.

So, we obtain $u_1 = \geq^{2x}$, and $u_2 = x\geq^{2x}$. The Wronskian of these two functions is

$$\begin{vmatrix} e^{-2x} & xe^{-2x} \\ -2e^{-2x} & -e^{-2x}(2x-1) \end{vmatrix} = -e^{-2x}e^{-2x}(2x-1) + 2xe^{-2x}e^{-2x}$$

$$= -e^{-4x}(2x-1) + 2xe^{-4x} = (-2x+1+2x)e^{-4x} = e^{-4x}.$$

Because the Wronskian is non-zero, the two functions are linearly independent, so this is in fact the general solution for the homogeneous differential equation (and not a mere subset of it).

We seek functions $A(x)$ and $B(x)$ so $A(x)u_1 + B(x)u_2$ is a general solution of the non-homogeneous equation. We need only calculate the integrals

$$A(x) = -\int \frac{1}{W} u_2(x) b(x) dx,\ B(x) = \int \frac{1}{W} u_1(x) b(x) dx$$

that is,

$$A(x) = -\int \frac{1}{e^{-4x}} xe^{-2x} \cosh x dx = -\int xe^{2x} \cosh x dx$$

$$= -\frac{1}{18} e^{x}(9(x-1) + e^{2x}(3x-1)) + C_1$$

$$B(x) = \int \frac{1}{e^{-4x}} e^{-2x} \cosh x dx = \int e^{2x} \cosh x dx$$

$$= \frac{1}{6} e^{x}(3 + e^{2x}) + C_2$$

where C_1 and C_2 are constants of integration.

Separable Partial Differential Equation

A separable partial differential equation (PDE) is one that can be broken into a set of separate equations of lower dimensionality (fewer independent variables) by a method of separation of variables.

This generally relies upon the problem having some special form or symmetry. In this way, the PDE can be solved by solving a set of simpler PDEs, or even ordinary differential equations (ODEs) if the problem can be broken down into one-dimensional equations.

The most common form of separation of variables is simple separation of variables in which a solution is obtained by assuming a solution of the form given by a product of functions of each individual coordinate.

There is a special form of separation of variables called R-separation of variables which is accomplished by writing the solution as a particular fixed function of the coordinates multiplied by a product of functions of each individual coordinate.

Laplace's equation on $\mathbb{R}^n$ is an example of a partial differential equation which admits solutions through -separation of variables.

(This should not be confused with the case of a separable ODE, which refers to a somewhat different class of problems that can be broken into a pair of integrals)

Universal Differential Equation

A universal differential equation (UDE) is a non-trivial differential algebraic equation with the property that its solutions can approximate any continuous function on any interval of the real line to any desired level of accuracy.

Lienard Equation

In mathematics, more specifically in the study of dynamical systems and differential equations, a Liénard equation is a second order differential equation, named after the French physicist Alfred-Marie Liénard.

During the development of radio and vacuum tube technology, Liénard equations were intensely studied as they can be used to model oscillating circuits. Under certain additional assumptions Liénard's theorem guarantees the uniqueness and existence of a limit cycle for such a system.

Definition

Let f and g be two continuously differentiable functions on R, with g an odd function and f an even function. Then the second order ordinary differential equation of the form

$$\frac{d^2x}{dt^2} + f(x)\frac{dx}{dt} + g(x) = 0$$

is called the Liénard equation.

Lienard System

The equation can be transformed into an equivalent two-dimensional system of ordinary differential equations. We define

$$F(x) := \int_0^x f(\xi)d\xi$$

$$x_1 := x$$

$$x_2 := \frac{dx}{dt} + F(x)$$

then

$$\begin{bmatrix} \dot{x}_1 \\ \dot{x}_2 \end{bmatrix} = \mathbf{h}(x_1, x_2) := \begin{bmatrix} x_2 - F(x_1) \\ -g(x_1) \end{bmatrix}$$

is called a Liénard system.

Alternatively, since Liénard equation itself is also an autonomous differential equation, the substitution $v = \frac{dx}{dt}$ leads the Liénard equation to become a first order differential equation:

$$v\frac{dv}{dx} + f(x)v + g(x) = 0$$

which belongs to Abel equation of the second kind.

Example

The Van der Pol oscillator

$$\frac{d^2x}{dt^2} - \mu(1 - x^2)\frac{dx}{dt} + x = 0$$

is a Liénard equation.

Lienard's Theorem

A Liénard system has a unique and stable limit cycle surrounding the origin if it satisfies the following additional properties:

- $g(x) > 0$ for all $x > 0$;

- $\lim_{x\to\infty} F(x) := \lim_{x\to\infty} \int_0^x f(\xi)d\xi = \infty;$
- $F(x)$ has exactly one positive root at some value p, where $F(x) < 0$ for $0 < x < p$ and $F(x) > 0$ and monotonic for $x > p$.

Method of Matched Asymptotic Expansions

In mathematics, particularly in solving singularly perturbed differential equations, the method of matched asymptotic expansions is a common approach to finding an accurate approximation to a problem's solution.

Method Overview

In a large class of singularly perturbed problems, the domain may be divided into two subdomains. On one of these, the solution is accurately approximated by an asymptotic series found by treating the problem as a regular perturbation.

The other subdomain consists of one or more small areas in which that approximation is inaccurate, generally because the perturbation terms in the problem are not negligible there. These areas are referred to as transition layers, or boundary or interior layers depending on whether they occur at the domain boundary (as is the usual case in applications) or inside the domain.

An approximation in the form of an asymptotic series is obtained in the transition layer(s) by treating that part of the domain as a separate perturbation problem. This approximation is called the "inner solution," and the other is the "outer solution," named for their relationship to the transition layer(s).

The outer and inner solutions are then combined through a process called "matching" in such a way that an approximate solution for the whole domain is obtained.

Location of Boundary Layer

Conveniently, we can see that the boundary layer, where $y'$and y'' are large, is near $t = 0$, as we supposed earlier. If we had supposed it to be at the other endpoint and proceeded by making the rescaling $\tau = (1-t)/\epsilon$, we would have found it impossible to satisfy the resulting matching condition.

For many problems, this kind of trial and error is the only way to determine the true location of the boundary layer.

Malgrange–Ehrenpreis Theorem

In mathematics, the Malgrange–Ehrenpreis theorem states that every non-zero linear differential operator with constant coefficients has a Green's function. It was first proved independently by Leon Ehrenpreis (1954, 1955) and Bernard Malgrange (1955–1956).

This means that if P is a polynomial (in several variables) then the differential equation

$$P(\partial / \partial x_i)u(x) = \delta(x)$$

has a distributional solution u, where δ is the Dirac delta function. It can be used to show that

$$P(\partial / \partial x_i)u(x) = f(x)$$

has a solution for any distribution f. The solution is not unique in general.

The analogue for differential operators whose coefficients are polynomials (rather than constants) is false.

Proofs

The original proofs of Malgrange and Ehrenpreis were non-constructive as they used the Hahn–Banach theorem. Since then several constructive proofs have been found.

There is a very short proof using the Fourier transform and the Bernstein–Sato polynomial, as follows. By taking Fourier transforms the Malgrange–Ehrenpreis theorem is equivalent to the fact that every non-zero polynomial P has a distributional inverse.

By replacing P by the product with its complex conjugate, one can also assume that P is non-negative.

For non-negative polynomials P the existence of a distributional inverse follows from the existence of the Bernstein–Sato polynomial, which implies that P^s can be analytically continued as a meromorphic distribution-valued function of the complex variable s; the constant term of the Laurent expansion of P^s at $s = \bullet 1$ is then a distributional inverse of P.

Other proofs, often giving better bounds on the growth of a solution, are given in (Hörmander 1983a, Theorem 7.3.10), (Reed & Simon 1975, Theorem IX.23, p. 48) and (Rosay 1991). (Hörmander 1983b, chapter 10) gives a detailed discussion of the regularity properties of the fundamental solutions.

A short constructive proof was presented in (Wagner 2009, Proposition 1, p. 458):

$$E = \frac{1}{\overline{P_m(2\eta)}} \sum_{j=0}^{m} a_j e^{\lambda_j \eta x} \mathcal{F}_\xi^{-1}\left(\frac{\overline{P(i\xi + \lambda_j \eta)}}{P(i\xi + \lambda_j \eta)} \right)$$

is a fundamental solution of $P(\partial)$, i.e., $P(\partial)E = \delta$, if P_m is the principal part of P, $\eta \in \mathbb{R}^n$ with $P_m(\eta) \neq 0$, the real numbers $\lambda_0, \ldots, \lambda_m$ are pairwise different, and $a_j = \prod_{k=0, k\neq j}^{m} (\lambda_j - \lambda_k)^{-1}$.

Nahm equations

The Nahm equations are a system of ordinary differential equations introduced by Werner Nahm in the context of the Nahm transform – an alternative to Ward's twistor construction of monopoles. The Nahm equations are formally analogous to the algebraic equations in the ADHM construction of instantons, where finite order matrices are replaced by differential operators.

Deep study of the Nahm equations was carried out by Nigel Hitchin and Simon Donaldson. Conceptually, the equations arise in the process of infinite-dimensional hyperkähler reduction. Among their many applications we can mention:

Hitchin's construction of monopoles, where this approach is critical for establishing nonsingularity of monopole solutions; Donaldson's description of the moduli space of monopoles; and the existence of hyperkähler structure on coadjoint orbits of complex semisimple Lie groups, proved by Peter Kronheimer, Olivier Biquard, and A.G. Kovalev.

Equations

Let $T_1(z), T_2(z)$, $T_3(z)$ be three matrix-valued meromorphic functions of a complex variable z. The Nahm equations are a system of matrix differential equations

$$\frac{dT_1}{dz} = [T_2, T_3]$$

$$[3pt]\frac{dT_2}{dz} = [T_3, T_1]$$

$$[3pt]\frac{dT_3}{dz} = [T_1, T_2],$$

together with certain analyticity properties, reality conditions, and boundary conditions. The three equations can be written concisely using the Levi-Civita symbol, in the form

$$\frac{dT_i}{dz} = \frac{1}{2}\sum_{j,k}\epsilon_{ijk}[T_j, T_k] = \sum_{j,k}\epsilon_{ijk}T_jT_k.$$

More generally, instead of considering N by N matrices, one can consider Nahm's equations with values in a Lie algebra g.

Additional Conditions

The variable z is restricted to the open interval (0,2), and the following conditions are imposed:

1. $T_i^* = -T_i$;
2. $T_i(2-z) = T_i(z)^T$;
3. T_i can be continued to a meromorphic function of z in a neighbourhood of the closed interval [0,2], analytic outside of 0 and 2, and with simple poles at $z = 0$ and $z = 2$; and
4. At the poles, the residues of (T_1, T_2, T_3) form an irreducible representation of the group SU(2).

Nahm–Hitchin Description of Monopoles

There is a natural equivalence between

1. the monopoles of charge k for the group SU(2), modulo gauge transformations, and
2. the solutions of Nahm equations satisfying the additional conditions above, modulo the simultaneous conjugation of T_1, T_2, T_3 by the group O(k,R).

Lax Representation

The Nahm equations can be written in the Lax form as follows.

Set

$$A_0 = T_1 + iT_2, \quad A_1 = -2iT_3, \quad A_2 = T_1 - iT_2$$
$$[3pt] \quad A(\zeta) = A_0 + \zeta A_1 + \zeta^2 A_2, \quad B(\zeta) = \frac{1}{2}\frac{dA}{d\zeta} = \frac{1}{2}A_1 + \zeta A_2,$$

then the system of Nahm equations is equivalent to the Lax equation

$$\frac{dA}{dz} = [A, B].$$

As an immediate corollary, we obtain that the spectrum of the matrix A does not depend on z. Therefore, the characteristic equation

$$\det(\lambda I + A(\zeta, z)) = 0,$$

which determines the so-called spectral curve in the twistor space TP^1, is invariant under the flow in z.

Chapter 4

Analytic Element Method

The analytic element method (AEM) is a numerical method used for the solution of partial differential equations. It was initially developed by O.D.L. Strack at the University of Minnesota. It is similar in nature to the boundary element method (BEM), as it does not rely upon discretisation of volumes or areas in the modelled system; only internal and external boundaries are discretised. One of the primary distinctions between AEM and BEMs is that the boundary integrals are calculated analytically.

The analytic element method is most often applied to problems of groundwater flow governed by the Poisson equation, though it is applicable to a variety of linear partial differential equations, including the Laplace, Helmholtz, and biharmonic equations.

The basic premise of the analytic element method is that, for linear differential equations, elementary solutions may be superimposed to obtain more complex solutions. A suite of 2D and 3D analytic solutions ("elements") are available for different governing equations. These elements typically correspond to a discontinuity in the dependent variable or its gradient along a geometric boundary (e.g., point, line, ellipse, circle, sphere, etc.). This discontinuity has a specific functional form (usually a polynomial in 2D) and may be manipulated to satisfy Dirichlet, Neumann, or Robin (mixed) boundary conditions. Each analytic solution is infinite in space and/or time. In addition, each analytic solution contains degrees of freedom (coefficients) that may be calculated to meet prescribed boundary conditions along the element's border. To obtain a global solution (i.e., the correct element coefficients), a system of equations is solved such that the boundary conditions are satisfied along all of the elements (using collocation, least-squares minimisation, or a similar approach). Notably, the global solution provides a spatially continuous description of the dependent variable everywhere in the infinite domain, and the governing equation is satisfied everywhere exactly except along the border of the element, where the governing equation is not strictly applicable due to the discontinuity.

Beeman's Algorithm

Beeman's algorithm is a method for numerically integrating ordinary differential equations of order 2, more specifically Newton's equations of motion $\ddot{x} = A(x)$. It was designed to allow high numbers of particles in simulations of molecular dynamics. There is a direct or explicit and an implicit variant of the method.

The direct variant was published by Schofield in 1973 as personal communication by Beeman. This is what is commonly known as Beeman's method. It is a variant of the Verlet integration method. It produces identical positions, but uses a different formula for the velocities. Beeman in 1976 published a class of implicit (predictor-corrector) multi-step methods, where Beeman's method is the direct variant of the third order method in this class.

Equation

The formula used to compute the positions at time $t+\Delta t$ in the full predictor-corrector scheme is:

- Predict $x(t+\Delta t)$ from data at times t and $t-\Delta t$
$$x(t+\Delta t) = x(t) + v(t)\Delta t + \frac{1}{6}\big(4a(t) - a(t-\Delta t)\big)\Delta t^2 + O(\Delta t^4)$$
- Correct position and velocities at time $t+\Delta t$ from data at times t and $t+\Delta t$ by repeated evaluation of the differential equation to get the acceleration $a(t+\Delta t)$ and of the equations of the implicit system

$$x(t+\Delta t) \quad = x(t) + v(t)\Delta t + \frac{1}{6}\big(a(t+\Delta t) + 2a(t)\big)\Delta t^2 + O(\Delta t^4);$$

$$v(t+\Delta t)\Delta t \quad = x(t+\Delta t) - x(t) + \frac{1}{6}\big(2a(t+\Delta t) + a(t)\big)\Delta t^2 + O(\Delta t^4);$$

In tests it was found that this corrector step needs to be repeated at most twice. The values on the right are the old values of the last iterations, resulting in the new values on the left.

Using only the predictor formula and the corrector for the velocities one obtains a direct or explicit method which is a variant of the Verlet integration method:

$$x(t+\Delta t) = x(t) + v(t)\Delta t + \frac{1}{6}\big(4a(t) - a(t-\Delta t)\big)\Delta t^2 + O(\Delta t^4)$$

$$v(t+\Delta t) = v(t) + \frac{1}{6}\big(2a(t+\Delta t) + 5a(t) - a(t-\Delta t)\big)\Delta t + O(\Delta t^3);$$

This is the variant that is usually understood as *Beeman's method.*

Beeman also proposed to alternatively replace the velocity update in the last equation by the second order Adams–Moulton method:

$$v(t+\Delta t) = v(t) + \frac{1}{12}\big(5a(t+\Delta t) + 8a(t) - a(t-\Delta t)\big)\Delta t + O(\Delta t^3)$$

where

- t is present time (i.e.: independent variable)
- Δt is the time step size
- $x(t)$ is the position at time t
- $v(t)$ is the velocity at time t
- $a(t)$ is the acceleration at time t, computed as a function of $x(t)$
- the last term is the error term, using the big O notation

Memory Requirements

The simulation must keep track of position, velocity, acceleration and previous acceleration vectors per particle (though some clever work-arounds for storing the previous acceleration vector are possible), keeping its memory requirements on par with velocity Verlet and slightly more expensive than the original Verlet method.

Bi-directional Delay Line

In mathematics, a bi-directional delay line is a numerical analysis technique used in computer simulation for solving ordinary differential equations by converting them to hyperbolic equations. In this way an explicit solution scheme is obtained with highly robust numerical properties. It was introduced by Auslander in 1968.

It originates from simulation of hydraulic pipelines where wave propagation was studied. It was then found that it could be used as an efficient numerical technique for numerically insulating different parts of a simulation model in each times step. It is used in the HOPSAN simulation package (Krus *et al.* 1990).

It is also known as the *transmission line method* TLM from an independent development by Johns and O'Brian 1980. This is also extended to partial differential equations.

Boundary Element Method

The boundary element method (BEM) is a numerical computational method of solving linear partial differential equations which have been formulated as integral equations (i.e. in *boundary integral* form). It can be applied in many areas of engineering and science including fluid mechanics, acoustics, electromagnetics, and fracture mechanics. (In electromagnetics, the more traditional term "method of moments" is often, though not always, synonymous with "boundary element method".)

Mathematical Basis

The integral equation may be regarded as an exact solution of the governing partial differential equation. The boundary element method attempts to use the given boundary conditions to fit boundary values into the integral equation, rather than values throughout the space defined by a partial differential equation. Once this is done, in the post-processing stage, the integral equation can then be used again to calculate numerically the solution directly at any desired point in the interior of the solution domain.

BEM is applicable to problems for which Green's functions can be calculated. These usually involve fields in linear homogeneous media. This places considerable restrictions on the range and generality of problems to which boundary elements can usefully be applied. Nonlinearities can be included in the formulation, although they will generally introduce volume integrals which then require the volume to be discretised before solution can be attempted, removing one of the most often cited advantages of BEM. A useful technique for treating the volume integral without discretising the volume is the dual-reciprocity method. The technique approximates part of the integrand using radial basis functions (local interpolating functions) and converts the volume integral into boundary integral after collocating at selected points distributed throughout the volume domain (including the boundary). In the dual-reciprocity BEM, although there is no need to discretise the volume into meshes, unknowns at chosen points inside the solution domain are involved in the linear algebraic equations approximating the problem being considered.

The Green's function elements connecting pairs of source and field patches defined by the mesh form a matrix, which is solved numerically. Unless the Green's function is well behaved, at least for pairs of patches near each other, the Green's function must be integrated over either or both the source patch and the field patch. The form of the method in which the integrals over the source and field patches are the same is called "Galerkin's method". Galerkin's method is the obvious approach for problems which are symmetrical with respect to exchanging the source and field points.

In frequency domain electromagnetics, this is assured by electromagnetic reciprocity. The cost of computation involved in naive Galerkin implementations is typically quite severe. One must loop over elements twice (so we get n^2 passes through) and for each pair of elements we loop through Gauss points in the elements producing a multiplicative factor proportional to the number of Gauss-points squared. Also, the function evaluations required are typically quite expensive, involving trigonometric/hyperbolic function calls. Nonetheless, the principal source of the computational cost is this double-loop over elements producing a fully populated matrix.

The Green's functions, or fundamental solutions, are often problematic to integrate as they are based on a solution of the system equations subject to a singularity load (e.g. the electrical field arising from a point charge). Integrating such singular fields is not easy. For simple element geometries (e.g. planar triangles) analytical integration can be used. For more general elements, it is possible to design purely numerical schemes that adapt to the singularity, but at great computational cost. Of course, when source point and target element (where the integration is done) are far-apart, the local gradient surrounding the point need not be quantified exactly and it becomes possible to integrate easily due to the smooth decay of the fundamental solution. It is this feature that is typically employed in schemes designed to accelerate boundary element problem calculations.

Comparison to Other Methods

The boundary element method is often more efficient than other methods, including finite elements, in terms of computational resources for problems where there is a small surface/volume ratio. Conceptually, it works by constructing a "mesh" over the modelled surface. However, for many problems boundary element methods are significantly less efficient than volume-discretisation methods (finite element method, finite difference method, finite volume method).

Boundary element formulations typically give rise to fully populated matrices. This means that the storage requirements and computational time will tend to grow according to the square of the problem size. By contrast, finite element matrices are typically banded (elements are only locally connected) and the storage requirements for the system matrices typically grow quite linearly with the problem size. Compression techniques (e.g. multipole expansions or adaptive cross approximation/hierarchical matrices) can be used to ameliorate these problems, though at the cost of added complexity and with a success-rate that depends heavily on the nature of the problem being solved and the geometry involved.

Cell Lists

Cell lists (also sometimes referred to as Cell linked-lists) are a tool for finding all atom pairs within a given cut-off distance of each other in Molecular dynamics simulations. These pairs are needed to compute the short-range non-bonded interactions in a system, such as Van der Waals forces or the short-range part of the electrostatic interaction when using Ewald summation.

Algorithm

a)

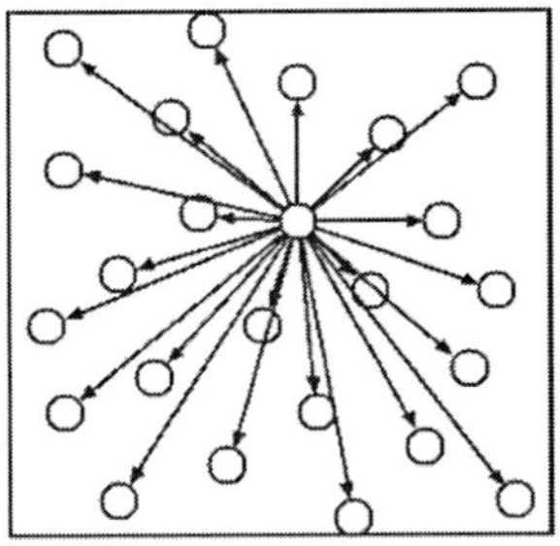

b)

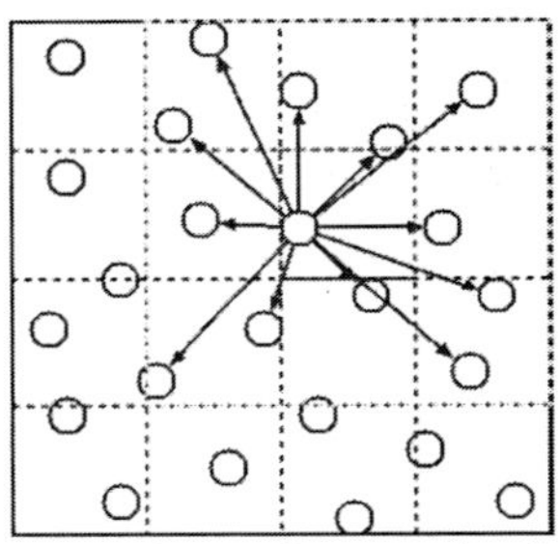

The pairwise interactions for a single particle can be computed by a) computing the interaction to all other particles or b) by dividing the domain into *cells* with an edge length of at least the cut-off radius of the interaction potential and computing the interaction between the particle and all particles in the same (red) and in the adjacent (green) cells.

Cell lists work by subdividing the simulation domain into cells with an edge length greater than or equal to the cut-off radius of the interaction to be computed. The particles are sorted into these cells and the interactions are computed between particles in the same or neighbouring cells.

In its most basic form, the non-bonded interactions for a cut-off distance r_c are computed as follows:

for all neighbouring cell pairs (C_α, C_β) do

for all $p_\alpha \in C_\alpha$ do

for all $p_\beta \in C_\beta$ do

$r^2 = \|\mathbf{x}[p_\alpha] - \mathbf{x}[p_\beta]\|_2^2$

if $r^2 \leq r_c^2$ then

Compute the interaction between p_a and p_β.

end if

end for

end for

end for

Since the cell length is at least r_c in all dimensions, no particles within of each other can be missed.

Given a simulation with N particles with a homogeneous particle density, the number of cells m is proportional to and inversely proportional to the cut-off radius (i.e. if increases, so does the number of cells).

The average number of particles per cell $\overline{c} = N / m$ therefore does not depend on the total number of particles. The cost of interacting two cells is in $\mathcal{O}(\overline{c}^2)$. The number of cell pairs is proportional to the number of cells which is again proportional to the number of particles N.

The total cost of finding all pairwise distances within a given cut-off is in $\mathcal{O}(Nc) \in \mathcal{O}(N)$, which is significantly better than computing the $\mathcal{O}(N^2)$ pairwise distances naively.

Periodic Boundary Conditions

In most simulations, Periodic boundary conditions are used to avoid imposing artificial boundary conditions. Using cell lists, these boundaries can be implemented in two ways

Ghost Cells

In the ghost cells approach, the simulation box is wrapped in an additional layer of cells. These cells contain periodically wrapped copies of the corresponding simulation cells inside the domain.

Although the data—and usually also the computational cost—is doubled for interactions over the periodic boundary, this approach has the advantage of being straightforward to implement and very easy to parallelise, since cells will only interact with their geographical neighbours.

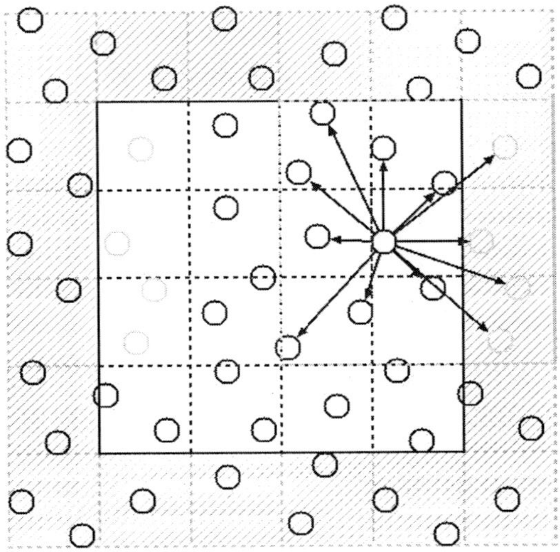

Figure: *Periodic boundary conditions can be simulating by wrapping the simulation box in an additional layer of cells (shaded orange) containing periodic copies of the boundary cells (blue particles)*

Periodic Wrapping

Instead of creating ghost cells, cell pairs that interact over a periodic boundary can also use a periodic correction vector $q_{\alpha\beta}$. This vector, which can be stored or computed for every cell pair (C_α, C_β) contains the correction which needs to be applied to "wrap" one cell around the domain to neighbour the other. The pairwise distance between two particles $p_\alpha \in C_\alpha$ and $p_\beta \in C_\beta$ is then computed as

$$r^2 = \left\| x[p_\alpha] - x[p_\beta] - q_{\alpha\beta} \right\|_2^2$$

This approach, although more efficient than using ghost cells, is less straightforward to implement (the cell pairs need to be identified over the periodic boundaries and the vector $q_{\alpha\beta}$ needs to be computed/stored).

Improvements

Despite reducing the computational cost of finding all pairs within a given cut-off distance from $\mathcal{O}(N^2)$ to $\mathcal{O}(N)$, the cell list algorithm listed above still has some inefficiencies.

Consider a computational cell with edge length equal to the cut-off radius r_c. The pairwise distance between all particles in the cell and in one of the neighbouring cells is computed. The cell has 26 neighbours: 6 sharing a common face, 12 sharing a common edge and 8 sharing a common corner. Of all the pairwise distances computed, only about 16% percent will actually be less than or equal to. In other words, 84% of all pairwise distance computations are spurious.

One way of overcoming this inefficiency is to partition the domain into cells of edge length smaller than. The pairwise interactions are then not just computed between neighbouring cells, but between all cells within of each other (first suggested in and implemented and analysed in, and). This approach can be taken to the limit wherein each cell holds at most one single particle, therefore reducing the number of spurious pairwise distance evaluations to zero. This gain in efficiency, however, is quickly offset by the number of cells that need to be inspected for every interaction with a cell, which, for example in 3d, grows cubically with the inverse of the cell edge length. Setting the edge length to, however, already reduces the number of spurious distance evaluations to 63%.

Another approach is outlined and tested in, in which the particles are first sorted along the axis connecting the cell centres. This approach generates only about 40% spurious pairwise distance computations, yet carries an additional cost due to sorting the particles.

Constraint Algorithm

In mechanics, a constraint algorithm is a method for satisfying constraints for bodies that obey Newton's equations of motion. There are three basic approaches to satisfying such constraints: choosing novel unconstrained coordinates ("internal coordinates"), introducing explicit constraint forces, and minimising constraint forces implicitly by the technique of Lagrange multipliers or projection methods.

Constraint algorithms are often applied to molecular dynamics simulations. Although such simulations are sometimes carried out in internal coordinates that automatically satisfy the bond-length and bond-angle constraints, they may also be carried out with explicit or implicit constraint forces for the bond lengths and bond angles. Explicit constraint forces typically shorten the time-step significantly, making the simulation less efficient computationally; in other words, more computer power is required to compute a trajectory of a given length. Therefore, internal coordinates and implicit-force constraint solvers are generally preferred.

Mathematical Background

The motion of a set of *N* particles can be described by a set of second-order ordinary differential equations, Newton's second law, which can be written in matrix form

$$M \cdot \frac{d^2 q}{dt^2} = f = -\frac{\partial V}{\partial q}$$

where M is a *mass matrix* and q is the vector of generalised coordinates that describe the particles' positions. For example, the vector q may be a *3N* Cartesian coordinates of the particle positions r_k, where *k* runs from 1 to *N*; in the absence of constraints, M would be the *3N*x*3N* diagonal square matrix of the particle masses. The vector f represents the generalised forces and the scalar *V*(q) represents the potential energy, both of which are functions of the generalised coordinates q.

If *M* constraints are present, the coordinates must also satisfy *M* time-independent algebraic equations

$$g_j(q) = 0$$

where the index *j* runs from 1 to *M*. For brevity, these functions g_i are grouped into an *M*-dimensional vector g below. The task is to solve the combined set of differential-algebraic (DAE) equations, instead of just the ordinary differential equations (ODE) of Newton's second law.

This problem was studied in detail by Joseph Louis Lagrange, who laid out most of the methods for solving it. The simplest approach is to define new generalised coordinates that are unconstrained; this approach eliminates the algebraic equations and reduces the problem once again to solving an ordinary differential equation. Such an approach is used, for example, in describing the motion of a rigid body; the position and orientation of a rigid body can be described by six independent, unconstrained coordinates, rather than describing the positions of the particles that make it up and the constraints among them that maintain their relative distances.

The drawback of this approach is that the equations may become unwieldy and complex; for example, the mass matrix M may become non-diagonal and depend on the generalised coordinates.

A second approach is to introduce explicit forces that work to maintain the constraint; for example, one could introduce strong spring forces that enforce the distances among mass points within a "rigid" body. The two difficulties of this approach are that the constraints are not satisfied exactly, and the strong forces may require very short time-steps, making simulations inefficient computationally.

A third approach is to use a method such as Lagrange multipliers or projection to the constraint manifold to determine the coordinate adjustments necessary to satisfy the constraints. Finally, there are various hybrid approaches in which different sets of constraints are satisfied by different methods, e.g., internal coordinates, explicit forces and implicit-force solutions.

Internal Coordinate Methods

The simplest approach to satisfying constraints in energy minimisation and molecular dynamics is to represent the mechanical system in so-called *internal coordinates* corresponding to unconstrained independent degrees of freedom of the system. For example, the dihedral angles of a protein are an independent set of coordinates that specify the positions of all the atoms without requiring any constraints.

The difficulty of such internal-coordinate approaches is twofold: the Newtonian equations of motion become much more complex and the internal coordinates may be difficult to define for cyclic systems of constraints, e.g., in ring puckering or when a protein has a disulfide bond.

The original methods for efficient recursive energy minimisation in internal coordinates were developed by Gτ and coworkers. Efficient recursive, internal-coordinate constraint solvers were extended to molecular dynamics. Analogous methods were applied later to other systems.

Hybrid Methods

Hybrid methods have also been introduced in which the constraints are divided into two groups; the constraints of the first group are solved using internal coordinates whereas those of the second group are solved using constraint forces, e.g., by a Lagrange multiplier or projection method. This approach was pioneered by Lagrange, and result in *Lagrange equations of the mixed type*.

Diffuse Element Method

The diffuse element method (DEM) or simply Diffuse Approximation is a computer simulation technique used in engineering analysis. It is a meshfree method. The diffuse element method was developed by B. Nayroles, G. Touzot and Pierre Villon at the Universite de Technologie de Compiegne, in 1992. It is in concept rather similar to the much older smoothed particle hydrodynamics. In the paper they describe a "diffuse approximation method", a method for function approximation from a given set of points.

In fact the method boils down to the well-known moving least squares for the particular case of a global approximation (using all available data points). Using this function approximation method, partial differential equations and thus fluid dynamic problems can be solved. For this, they coined the term Diffuse Element Method (DEM). Advantages over finite element methods are that DEM doesn't rely on a grid, and is more precise in the evaluation of the derivatives of the reconstructed functions.

Direct Multiple Shooting Method

In the area of mathematics known as numerical ordinary differential equations, the direct multiple shooting method is a numerical method for the solution of boundary value problems. The method divides the interval over which a solution is sought into several smaller intervals, solves an initial value problem in each of the smaller intervals, and imposes additional matching conditions to form a solution on the whole interval. The method constitutes a significant improvement in distribution of nonlinearity and numerical stability over single shooting methods.

Single Shooting Methods

Shooting methods can be used to solve boundary value problems (BVP) like

$$y'(t) = f(t, y(t)), \quad y(t_a) = y_a, \quad y(t_b) = y_b,$$

in which the time points t_a and t_b are known but the initial and terminal values y_a and y_b are unknown and sought for.

Single shooting methods proceed as follows. Let $y(t;\, t_0,\, y_0)$ denote the solution of the initial value problem (IVP)

$$y'(t) = f(t, y(t)), \quad y(t_0) = y_0.$$

Define the function $F(p)$ as the difference between $y(t_b;\, p)$ and the specified boundary value y_b: $F(p) = y(t_b;\, p) - y_b$. Then for every solution $(y_a,\, y_b)$ of the boundary value problem we have $y_a{=}y_0$ while y_b corresponds to a root of F. This root be solved by any root-finding method given that certain method-dependent prerequisites are satisfied. This often will require initial guesses to y_a and y_b. Typically, analytic root finding is impossible and iterative methods such as Newton's method are used for this task.

The application of single shooting for the numerical solution of boundary value problems suffers from several drawbacks.

- For a given initial value y_0 the solution of the IVP obviously must exist on the interval $[t_a{,}t_b]$ so that we can evaluate the function F whose root is sought.

For highly nonlinear or unstable ODEs, this requires the initial guess y_0 to be extremely close to an actual but unknown solution y_a.

Initial values that are chosen slightly off the true solution may lead to singularities or breakdown of the ODE solver method. Choosing such solutions is inevitable in an iterative root-finding method, however.

- Finite precision numerics may make it impossible at all to find initial values that allow for the solution of the ODE on the whole time interval.
- The nonlinearity of the ODE effectively becomes a nonlinearity of F, and requires a root-finding technique capable of solving nonlinear systems. Such methods typically converge slower as nonlinearities become more

severe. The boundary value problem solver's performance suffers from this.

- Even stable and well-conditioned ODEs may make for unstable and ill-conditioned BVPs. A slight alteration of the initial value guess y_0 may generate an extremely large step in the ODEs solution $y(t_b; t_a, y_0)$ and thus in the values of the function F whose root is sought. Non-analytic root-finding methods can seldom cope with this behaviour.

Multiple Shooting

A direct multiple shooting method partitions the interval $[t_a, t_b]$ by introducing additional grid points

$$t_a = t_0 < t_1 < \cdots < t_N = t_b$$

The method starts by guessing somehow the values of y at all grid points t_k with 0 d– k d– N – 1. Denote these guesses by y_k. Let $y(t; t_k, y_k)$ denote the solution emanating from the kth grid point, that is, the solution of the initial value problem

$$y'(t) = f(t, y(t)), \quad y(t_k) = y_k.$$

All these solutions can be pieced together to form a continuous trajectory if the values y match at the grid points. Thus, solutions of the boundary value problem correspond to solutions of the following system of N equations:

$$\begin{aligned} y(t_1; t_0, y_0) &= y_1 \\ &\vdots \\ y(t_{N-1}; t_{N-2}, y_{N-2}) &= y_{N-1} \\ y(t_N; t_{N-1}, y_{N-1}) &= y_b. \end{aligned}$$

The central N–2 equations are the matching conditions, and the first and last equations are the conditions $y(t_a) = y_a$ and $y(t_b) = y_b$ from the boundary value problem.

The multiple shooting method solves the boundary value problem by solving this system of equations. Typically, a modification of the Newton's method is used for the latter task.

Discontinuous Galerkin Method

Discontinuous Galerkin methods (DG methods) in mathematics form a class of numerical methods for solving differential equations. They combine features of the finite element and the finite volume framework and have been successfully applied to hyperbolic, elliptic, parabolic and mixed form problems arising from a wide range of applications.

DG methods have in particular received considerable interest for problems with a dominant first-order part, e.g. in electrodynamics, fluid mechanics and plasma physics.

Discontinuous Galerkin methods were first proposed and analysed in the early 1970s as a technique to numerically solve partial differential equations. In 1973 Reed and Hill introduced a DG method to solve the hyperbolic neutron transport equation.

The origin of the DG method for elliptic problems cannot be traced back to a single publication as features such as jump penalisation in the modern sense were developed gradually. However, among the early influential contributors were Babuška, J.-L. Lions, Nitsche and Zlamal. Interestingly DG methods for elliptic problems were already developed in a paper by Baker in the setting of 4th order equations in 1977.

A more complete account of the historical development and an introduction to DG methods for elliptic problems is given in a publication by Arnold, Brezzi, Cockburn and Marini. A number of research directions and challenges on DG methods are collected in the proceedings volume edited by Cockburn, Karniadakis and Shu.

Overview

Much like a the continuous Galerkin (CG) method, the discontinuous Galerkin (DG) method is a finite element method formulated relative to a weak formulation of a particular model system.

Unlike traditional CG methods that are conforming, the DG method works over a trial function space that is piecewise discontinuous, and thus often comprise more inclusive function spaces than the finite dimensional inner product subspaces utilised in conforming methods.

As an example, consider the continuity equation arising in electromagnetic theory:

$$\frac{\partial \rho}{\partial t} + \nabla \cdot \mathbf{J} = 0.$$

Now consider the finite dimensional space of discontinuous piecewise polynomial functions over the spatial domain Ω restricted to a discrete triangulation Ω_h, written as

$$S_h^p(\Omega_h) = \{v_{|\Omega_{e_i}} \in P^p(\Omega_{e_i}),\ \forall \Omega_{e_i} \in \Omega_h\}$$

for $P^p(\Omega_{e_i})$ the space of degree $\leq p$ polynomials over element Ω_{e_i} indexed by i. Then for finite element shape functions $N_j \in P^p$ the solution is represented by

$$\rho_h = \sum_{j=1}^{dofs} \rho_j^i(t) N_j^i(\mathbf{x}),\quad \forall \mathbf{x} \in \Omega_{e_i}.$$

Then similarly choosing a test function

$$\phi_h(\mathbf{x}) = \sum_{j=1}^{dofs} \phi_j^i N_l^i(\mathbf{x}),\quad \forall \mathbf{x} \in \Omega_{e_i},$$

multiplying the continuity equation by ϕ_h and integrating by parts in space, the semidiscrete DG formulation becomes:

$$\frac{d}{dt}\int_{\Omega_{e_i}} \rho_h \phi_h d\mathbf{x} + \int_{\partial\Omega_{e_i}} \phi_h \mathbf{J}_h \cdot \mathbf{n} d\mathbf{x} = \int_{\Omega_{e_i}} \mathbf{J}_h \cdot \nabla \phi_h d\mathbf{x}.$$

Discrete Element Method

A discrete element method (DEM), also called a distinct element method is any of family of numerical methods for computing the motion of a large number of particles of micrometre-scale size and above. Though DEM is very closely related to molecular dynamics, the method is generally distinguished by its inclusion of rotational degrees-of-freedom as well as stateful contact and often complicated geometries (including polyhedra). With advances in computing power and numerical algorithms for nearest neighbour sorting, it has become possible to numerically simulate millions of particles on a single processor. Today DEM is becoming widely accepted as an effective method of addressing engineering problems in granular and discontinuous materials, especially in granular flows, powder mechanics, and rock mechanics.

Discrete element methods are relatively computationally intensive, which limits either the length of a simulation or the number of particles. Several DEM codes, as do molecular dynamics codes, take advantage of parallel processing capabilities (shared or distributed systems) to scale up the number of particles or length of the simulation. An alternative to treating all particles separately is to average the physics across many particles and thereby treat the material as a continuum.

In the case of solid-like granular behaviour as in soil mechanics, the continuum approach usually treats the material as elastic or elasto-plastic and models it with the finite element method or a mesh free method. In the case of liquid-like or gas-like granular flow, the continuum approach may treat the material as a fluid and use computational fluid dynamics. Drawbacks to homogenisation of the granular scale physics, however, are well-documented and should be considered carefully before attempting to use a continuum approach.

The DEM Family

The various branches of the DEM family are the distinct element method proposed by Cundall in 1971, the generalised discrete element method proposed by Hocking, Williams and Mustoe in 1985, the discontinuous deformation analysis (DDA) proposed by Shi in 1988 and the finite-discrete element method concurrently developed by several groups (e.g., Munjiza and Owen). The general method was originally developed by Cundall in 1971 to problems in rock mechanics.

The theoretical basis of the method was established by Sir Isaac Newton in 1697. Williams, Hocking, and Mustoe in 1985 showed that DEM could be viewed as a generalized finite element method. Its application to geomechanics problems is described in the book *Numerical Modelling in Rock Mechanics*, by Pande, G., Beer, G. and Williams, J.R.. The 1st, 2nd and 3rd International Conferences on Discrete Element Methods have been a common point for researchers to publish advances in the method and its applications. Journal articles reviewing the state of the art have been published by Williams, Bicanic, and Bobet et al.

A comprehensive treatment of the combined Finite Element-Discrete Element Method is contained in the book *The Combined Finite-Discrete Element Method* by Munjiza.

Applications

The fundamental assumption of the method is that the material consists of separate, discrete particles. These particles may have different shapes and properties. Some examples are:

- liquids and solutions, for instance of sugar or proteins;
- bulk materials in storage silos, like cereal;
- granular matter, like sand;
- powders, like toner.
- Blocky or jointed rock masses

Typical industries using DEM are:

- Agriculture and food handling
- Chemical
- Civil Engineering
- Oil and gas
- Mining
- Mineral processing
- Pharmaceutical
- Powder metallurgy

Outline of the Method

A DEM-simulation is started by first generating a model, which results in spatially orienting all particles and assigning an initial velocity. The forces which act on each particle are computed from the initial data and the relevant physical laws and contact models. Generally, a simulation consists of three parts: the initialisation, explicit time-stepping, and post-processing. The time-stepping usually requires a nearest neighbour sorting step to reduce the number of possible contact pairs and decrease the computational requirements; this is often only performed periodically. The following forces may have to be considered in macroscopic simulations:

- friction, when two particles touch each other;
- contact plasticity, or recoil, when two particles collide;
- gravity, the force of attraction between particles due to their mass, which is only relevant in astronomical simulations.

- attractive potentials, such as cohesion, adhesion, liquid bridging, electrostatic attraction. Note that, because of the overhead from determining nearest neighbour pairs, exact resolution of long-range, compared with particle size, forces can increase computational cost or require specialised algorithms to resolve these interactions.

On a molecular level, we may consider

- the Coulomb force, the electrostatic attraction or repulsion of particles carrying electric charge;
- Pauli repulsion, when two atoms approach each other closely;
- van der Waals force.

All these forces are added up to find the total force acting on each particle. An integration method is employed to compute the change in the position and the velocity of each particle during a certain time step from Newton's laws of motion. Then, the new positions are used to compute the forces during the next step, and this loop is repeated until the simulation ends.

Typical integration methods used in a discrete element method are:

- the Verlet algorithm,
- velocity Verlet,
- symplectic integrators,
- the leapfrog method.

Long-range Forces

When long-range forces (typically gravity or the Coulomb force) are taken into account, then the interaction between each pair of particles needs to be computed. The number of interactions, and with it the cost of the computation, increases quadratically with the number of particles. This is not acceptable for simulations with large number of particles. A possible way to avoid this problem is to combine some particles, which are far away from the particle under consideration, into one pseudoparticle. Consider as an example the interaction between a star and a distant galaxy:

The error arising from combining all the stars in the distant galaxy into one point mass is negligible. So-called tree algorithms are used to decide which particles can be combined into one pseudoparticle. These algorithms arrange all particles in a tree, a quadtree in the two-dimensional case and an octree in the three-dimensional case.

However, simulations in molecular dynamics divide the space in which the simulation take place into cells. Particles leaving through one side of a cell are simply inserted at the other side (periodic boundary conditions); the same goes for the forces.

The force is no longer taken into account after the so-called cut-off distance (usually half the length of a cell), so that a particle is not influenced by the mirror image of the same particle in the other side of the cell. One can now increase the number of particles by simply copying the cells.

Algorithms to deal with long-range force include:

- Barnes–Hut simulation,
- the fast multipole method.

Combined Finite-discrete Element Method

Following the work by Munjiza and Owen's earlier work, the combined-discrete element method has been further developed to various irregular and deformable particles in many applications including pharmaceutical tableting, packaging and flow simulations, and concrete and impact analysis, and many other applications.

Advantages and Limitations

Advantages

- DEM can be used to simulate a wide variety of granular flow and rock mechanics situations. Several research groups have independently developed simulation software that agrees well with experimental findings in a wide range of engineering applications, including adhesive powders, granular flow, and jointed rock masses.
- DEM allows a more detailed study of the micro-dynamics of powder flows than is often possible using physical experiments. For example, the force networks formed in a granular media can be visualised using DEM. Such

measurements are nearly impossible in experiments with small and many particles.

Disadvantages

- The maximum number of particles, and duration of a virtual simulation is limited by computational power. Typical flows contain billions of particles, but contemporary DEM simulations on large cluster computing resources have only recently been able to approach this scale for sufficiently long time (simulated time, not actual program execution time).

Discrete Laplace Operator

In mathematics, the discrete Laplace operator is an analog of the continuous Laplace operator, defined so that it has meaning on a graph or a discrete grid. For the case of a finite-dimensional graph (having a finite number of edges and vertices), the discrete Laplace operator is more commonly called the Laplacian matrix.

The discrete Laplace operator occurs in physics problems such as the Ising model and loop quantum gravity, as well as in the study of discrete dynamical systems. It is also used in numerical analysis as a stand-in for the continuous Laplace operator. Common applications include image processing, where it is known as the Laplace filter, and in machine learning for clustering and semi-supervised learning on neighbourhood graphs.

Definitions

Graph Laplacians: There are various definitions of the *discrete Laplacian* for graphs, differing by sign and scale factor (sometimes one averages over the neighbouring vertices, other times one just sums; this makes no difference for a regular graph).

The traditional definition of the graph Laplacian, given below, corresponds to the negative continuous Laplacian on a domain with a free boundary.

Let $G=(V,E)$ be a graph with vertices v and edges E. Let $\phi : V \rightarrow R$ be a ring-valued function of the vertices. Then, the discrete Laplacian Δ acting on ϕ is defined by

$$(\Delta\phi)(v) = \sum_{w:d(w,v)=1} \left[\phi(w) - \phi(v)\right]$$

where $d(w,v)$ is the graph distance between vertices w and v. Thus, this sum is over the nearest neighbours of the vertex v. For a graph with a finite number of edges and vertices, this definition is identical to that of the Laplacian matrix. That is, ϕ can be written as a column vector; and so $\Delta\phi$ is the product of the column vector and the Laplacian matrix, while $(\Delta\phi)(v)$ is just the v'th entry of the product vector.

If the graph has weighted edges, that is, a weighting function $\gamma : E \rightarrow R$ is given, then the definition can be generalised to

$$(\Delta_\gamma\phi)(v) = \sum_{w:d(w,v)=1} \gamma_{wv} \left[\phi(w) - \phi(v)\right]$$

where γ_{wv} is the weight value on the edge $wv \in E$.

Closely related to the discrete Laplacian is the averaging operator:

$$(M\phi)(v) = \frac{1}{\deg v} \sum_{w:d(w,v)=1} \phi(w)$$

Mesh Laplacians: In addition to considering the connectivity of nodes and edges in a graph, mesh laplace operators take into account the geometry of a surface (e.g. the angles at the nodes). For triangle meshes, for example, different discretisations exist, some of them are an extension of the graph operator, while other approaches are based on the finite element method and allow for higher order approximations. An overview on some mesh operators and a comparison is given in.

Finite Differences: Approximations of the Laplacian, obtained by the finite difference method or by the finite element method can also be called Discrete Laplacians. For example, the Laplacian in two dimensions can be approximated using the five-point stencil finite difference method, resulting in

$$\Delta f(x,y) \approx \frac{f(x-h,y) + f(x+h,y) + f(x,y-h) + f(x,y+h) - 4f(x,y)}{h^2},$$

where the grid size is h in both dimensions, so that the five point stencil of a point (x, y) in the grid is

$$\{(x-h,y),(x,y),(x+h,y),(x,y-h),(x,y+h)\}.$$

If the grid size $h=1$, the result is the negative discrete Laplacian on the graph, which is the square lattice grid. There are no constraints here on the values of the function $f(x,y)$ on the boundary of the lattice grid, thus this is the case of the homogeneous Neumann boundary condition, i.e., free boundary. Other types of boundary conditions, e.g., the homogeneous Dirichlet boundary condition, where $f(x,y)=0$ on the boundary of the grid, are rarely used for graph Laplacians, but are common in other applications.

Multidimensional discrete Laplacians on rectangular cuboid regular grids have very special properties, e.g., they are Kronecker sums of one-dimensional discrete Laplacians, in which case all its eigenvalues and eigenvectors can be explicitly calculated.

Finite Element Method

In this approach, the domain is discretised into smaller elements, often triangles or tetrahedra, but other elements such as rectangles or cuboids are possible. The solution space is then approximated using so called form-functions of a pre-defined degree. The differential equation containing the Laplace operator is then transformed into a variational formulation, and a system of equations is constructed (linear or eigenvalue problems). The resulting matrices are usually very sparse and can be solved with iterative methods.

Spectrum

The spectrum of the discrete Laplacian is of key interest; since it is a self-adjoint operator, it has a real spectrum. For the convention $\Delta = I - M$, the spectrum lies within $[0,2]$ (as the averaging operator has spectral values in $[-1,1]$). The smallest non-zero eigenvalue is denoted λ_1 and is called the spectral gap. There is also the notion of the spectral radius, commonly taken as the largest eigenvalue.

The eigenvectors don't depend on the convention above (for regular graphs), and are the same as for the averaging operator (as they differ by adding a multiple of the identity), though the eigenvalues differ according to the convention.

For operators that approximate the underlying continuous Laplacian the eigenvalues are a sequence of positive real numbers. The first eigenvalue is zero, if the domain has a boundary and the Neumann boundary condition is used, or if the shape contains no boundary (e.g. the sphere).

Theorems

If the graph is an infinite square lattice grid, then this definition of the Laplacian can be shown to correspond to the continuous Laplacian in the limit of an infinitely fine grid. Thus, for example, on a one-dimensional grid we have

$$\frac{\partial^2 F}{\partial x^2} = \lim_{\epsilon \to 0} \frac{[F(x+\epsilon) - F(x)] + [F(x-\epsilon) - F(x)]}{\epsilon^2}.$$

This definition of the Laplacian is commonly used in numerical analysis and in image processing. In image processing, it is considered to be a type of digital filter, more specifically an edge filter, called the Laplace filter.

Discrete Schrödinger operator

Let $P : V \to R$ be a potential function defined on the graph. Note that P can be considered to be a multiplicative operator acting diagonally on ϕ

$$(P\phi)(v) = P(v)\phi(v).$$

Then $H = \Delta + P$ is the discrete Schrödinger operator, an analog of the continuous Schrödinger operator. If the number of edges meeting at a vertex is uniformly bounded, and the potential is bounded, then H is bounded and self-adjoint.

The spectral properties of this Hamiltonian can be studied with Stone's theorem; this is a consequence of the duality between posets and Boolean algebras.

On regular lattices, the operator typically has both travelling-wave as well as Anderson localisation solutions, depending on whether the potential is periodic or random.

ADE Classification

Certain equations involving the discrete Laplacian only have solutions on the simply-laced Dynkin diagrams (all edges multiplicity 1), and are an example of the ADE classification. Specifically, the only positive solutions to the homogeneous equation:

$$\Delta\phi = \phi,$$

in words,

"Twice any label is the sum of the labels on adjacent vertices,"

are on the extended (affine) ADE Dynkin diagrams, of which there are 2 infinite families (A and D) and 3 exceptions (E). The resulting numbering is unique up to scale, and if the smallest value is set at 1, the other numbers are integers, ranging up to 6.

The ordinary ADE graphs are the only graphs that admit a positive labelling with the following property:

Twice any label minus two is the sum of the labels on adjacent vertices.

In terms of the Laplacian, the positive solutions to the inhomogeneous equation:

$$\Delta\phi = \phi - 2.$$

The resulting numbering is unique (scale is specified by the "2"), and consists of integers; for E_8 they range from 58 to 270, and have been observed as early as (Bourbaki 1968).

Dynamic Errors of Numerical Methods of ODE Discretisation

The dynamical characteristic of the numerical method of ordinary differential equations (ODE) discretisation – is the natural logarithm of its function of stability $\mathbf{D} = \ln \rho(h\lambda)$. Dynamic characteristic is considered in three forms:

D– Complex dynamic characteristic;

D_R– Real dynamic characteristics;

D_I– Imaginary dynamic characteristics.

The dynamic characteristic represents the transformation operator of eigenvalues of a Jacobian matrix of the initial differential mathematical model (MM) in eigenvalues of a Jacobian matrix of mathematical model (also differential) whose exact solution passes through the discrete sequence of points of the initial MM solution received by given numerical method.

Chapter 5

Finite Element Method

The finite element method (FEM) (its practical application often known as finite element analysis (FEA)) is a numerical technique for finding approximate solutions to partial differential equations (PDE) and their systems, as well as (less often) integral equations. FEM is a special case of the more general Galerkin method with polynomial approximation functions.

The solution approach is based on eliminating the spatial derivatives from the PDE. This approximates the PDE with

- a system of algebraic equations for steady state problems,
- a system of ordinary differential equations for transient problems.

These equation systems are linear if the underlying PDE is linear, and vice versa. Algebraic equation systems are then solved using numerical linear algebra methods. Ordinary differential equations that arise in transient problems then numerically integrated using standard techniques such as Euler's method, Runge-Kutta, etc. There is a large number of open source and commercial finite element softwares. More recently, first web browser-based FEA applications became available such as in NCLab.

In solving partial differential equations, the primary challenge is to create an equation that approximates the equation to be studied, but is numerically stable, meaning that errors in the input and intermediate calculations do not accumulate and cause the resulting output to be meaningless. There are many ways of doing this, all with advantages and disadvantages.

The finite element method is a good choice for solving partial differential equations over complicated domains (like cars and oil pipelines), when the domain changes (as during a solid state reaction with a moving boundary), when the desired precision varies over the entire domain, or when the solution lacks smoothness.

For instance, in a frontal crash simulation it is possible to increase prediction accuracy in "important" areas like the front of the car and reduce it in its rear (thus reducing cost of the simulation). Another example would be in Numerical weather prediction, where it is more important to have accurate predictions over developing highly nonlinear phenomena (such as tropical cyclones in the atmosphere, or eddies in the ocean) rather than relatively calm areas.

History

The finite element method originated from the need for solving complex elasticity and structural analysis problems in civil and aeronautical engineering. Its development can be traced back to the work by Alexander Hrennikoff (1941) and Richard Courant (1942). While the approaches used by these pioneers are different, they share one essential characteristic: mesh discretisation of a continuous domain into a set of discrete sub-domains, usually called elements. Starting in 1947, Olgierd Zienkiewicz from Imperial College gathered those methods together into what would be called the Finite Element Method, building the pioneering mathematical formalism of the method.

Hrennikoff's work discretizes the domain by using a lattice analogy, while Courant's approach divides the domain into finite triangular subregions to solve second order elliptic partial differential equations (PDEs) that arise from the problem of torsion of a cylinder. Courant's contribution was evolutionary, drawing on a large body of earlier results for PDEs developed by Rayleigh, Ritz, and Galerkin. Development of the finite element method began in earnest in the middle to late 1950s for airframe and structural analysis and gathered momentum at the University of Stuttgart through the work of John Argyris and at Berkeley through the work of Ray W. Clough in the 1960s for use in civil engineering. By late 1950s, the key concepts of stiffness matrix and element assembly existed essentially in the form used today. NASA issued a request for proposals for the development of the finite element software NASTRAN in 1965.

The method was again provided with a rigorous mathematical foundation in 1973 with the publication of Strang and Fix's *An Analysis of The Finite Element Method*, and has since been generalised into a branch of applied mathematics for numerical modelling of physical systems in a wide variety of engineering disciplines, e.g., electromagnetism and fluid dynamics.

Technical Discussion

We will illustrate the finite element method using two sample problems from which the general method can be extrapolated. It is assumed that the reader is familiar with calculus and linear algebra.

Illustrative Problems P1 and P2

P1 is a one-dimensional problem

$$\text{P1 } :\begin{cases} u''(x) = f(x) \text{ in } (0,1), \\ u(0) = u(1) = 0, \end{cases}$$

where f is given, u is an unknown function of x, and u'' is the second derivative of with respect to.

P2 is a two-dimensional problem (Dirichlet problem)

$$\text{P2 } :\begin{cases} u_{xx}(x,y) + u_{yy}(x,y) = f(x,y) & \text{in } \Omega, \\ u = 0 & \text{on } \partial\Omega, \end{cases}$$

where Ω is a connected open region in the (x,y) plane whose boundary $\partial\Omega$ is "nice" (e.g., a smooth manifold or a polygon), and u_{xx} and u_{yy} denote the second derivatives with respect to x and y, respectively.

The problem P1 can be solved "directly" by computing antiderivatives. However, this method of solving the boundary value problem works only when there is only one spatial dimension and does not generalise to higher-dimensional problems or to problems like $u + u'' = f$. For this reason, we will develop the finite element method for P1 and outline its generalisation to P2.

Our explanation will proceed in two steps, which mirror two essential steps one must take to solve a boundary value problem (BVP) using the FEM.

- In the first step, one rephrases the original BVP in its weak form. Little to no computation is usually required for this step. The transformation is done by hand on paper.
- The second step is the discretisation, where the weak form is discretised in a finite dimensional space.

After this second step, we have concrete formulae for a large but finite dimensional linear problem whose solution will approximately solve the original BVP. This finite dimensional problem is then implemented on a computer.

Matrix form of the Problem

If we write $u(x)=\sum_{k=1}^{n}u_k v_k(x)$ and $f(x)=\sum_{k=1}^{n}f_k v_k(x)$ then problem (3), taking $v(x)=v_j(x)$ for $j=1,\dots,n$, becomes

$$-\sum_{k=1}^{n}u_k\phi(v_k,v_j)=\sum_{k=1}^{n}f_k\int v_k v_j dx \quad \text{for } j=1,\dots,n. \tag{4}$$

If we denote by u and f the column vectors $(u_1,\dots,u_n)^t$ and $(f_1,\dots,f_n)^t$, and if we let

$$L=(L_{ij})$$

and

$$M=(M_{ij})$$

be matrices whose entries are

$$L_{ij}=\phi(v_i,v_j)$$

and

$$M_{ij}=\int v_i v_j dx$$

then we may rephrase (4) as

$$-L\mathbf{u}=M\mathbf{f}\,. \tag{5}$$

It is not, in fact, necessary to assume $f(x)=\sum_{k=1}^{n}f_k v_k(x)$. For a general function $f(x)$, problem (3) with $v(x)=v_j(x)$ for $j=1,\dots,n$ becomes actually simpler, since no matrix M is used,

$$-Lu = b, \quad (6)$$

where $b = (b_1, \ldots, b_n)^t$ and $b_j = \int f v_j dx$ for $j = 1, \ldots, n$.

As we have discussed before, most of the entries of L and M are zero because the basis functions u_k have small support. So we now have to solve a linear system in the unknown u where most of the entries of the matrix, which we need to invert, are zero.

Such matrices are known as sparse matrices, and there are efficient solvers for such problems (much more efficient than actually inverting the matrix.) In addition, L is symmetric and positive definite, so a technique such as the conjugate gradient method is favoured. For problems that are not too large, sparse LU decompositions and Cholesky decompositions still work well. For instance, Matlab's backslash operator (which uses sparse LU, sparse Cholesky, and other factorisation methods) can be sufficient for meshes with a hundred thousand vertices.

The matrix is usually referred to as the stiffness matrix, while the matrix is dubbed the mass matrix.

General Form of the Finite Element Method

In general, the finite element method is characterised by the following process.

- One chooses a grid for Ω. In the preceding treatment, the grid consisted of triangles, but one can also use squares or curvilinear polygons.
- Then, one chooses basis functions. In our discussion, we used piecewise linear basis functions, but it is also common to use piecewise polynomial basis functions.

A separate consideration is the smoothness of the basis functions. For second order elliptic boundary value problems, piecewise polynomial basis function that are merely continuous suffice (i.e., the derivatives are discontinuous).

For higher order partial differential equations, one must use smoother basis functions. For instance, for a fourth order problem such as $u_{xxxx} + u_{yyyy} = f$, one may use piecewise quadratic basis functions that are C^1.

Another consideration is the relation of the finite dimensional space V to its infinite dimensional counterpart, in the examples above H_0^1.. A conforming element method is one in which the space is a subspace of the element space for the continuous problem. The example above is such a method. If this condition is not satisfied, we obtain a nonconforming element method, an example of which is the space of piecewise linear functions over the mesh which are continuous at each edge midpoint. Since these functions are in general discontinuous along the edges, this finite dimensional space is not a subspace of the original H_0^1..

Typically, one has an algorithm for taking a given mesh and subdividing it. If the main method for increasing precision is to subdivide the mesh, one has an h-method (h is customarily the diameter of the largest element in the mesh.) In this manner, if one shows that the error with a grid h is bounded above by Ch^p, for some $C < \infty$ and $p > 0$, then one has an order p method. Under certain hypotheses (for instance, if the domain is convex), a piecewise polynomial of order d method will have an error of order $p = d + 1$.

If instead of making h smaller, one increases the degree of the polynomials used in the basis function, one has a p-method. If one combines these two refinement types, one obtains an hp-method (hp-FEM). In the hp-FEM, the polynomial degrees can vary from element to element. High order methods with large uniform p are called spectral finite element methods (SFEM). These are not to be confused with spectral methods.

For vector partial differential equations, the basis functions may take values in $\mathbb{R}^n$.

Various Types of Finite Element Methods

AEM

The Applied Element Method, or AEM combines features of both FEM and Discrete element method, or (DEM).

Generalised Finite Element Method

The Generalised Finite Element Method (GFEM) uses local spaces consisting of functions, not necessarily polynomials, that reflect the available information on the unknown solution and thus ensure good local approximation. Then a partition of unity is used to "bond" these spaces together to form the approximating subspace. The effectiveness of GFEM has been shown when applied to problems with domains having complicated boundaries, problems with micro-scales, and problems with boundary layers.

hp-FEM

The hp-FEM combines adaptively elements with variable size h and polynomial degree p in order to achieve exceptionally fast, exponential convergence rates.

hpk-FEM

The hpk-FEM combines adaptively elements with variable size h, polynomial degree of the local approximations p and global differentiability of the local approximations $(k-1)$ in order to achieve best convergence rates.

Comparison to the Finite Difference Method

The finite difference method (FDM) is an alternative way of approximating solutions of PDEs. The differences between FEM and FDM are:

- The most attractive feature of the FEM is its ability to handle complicated geometries (and boundaries) with relative ease. While FDM in its basic form is restricted to handle rectangular shapes and simple alterations thereof, the handling of geometries in FEM is theoretically straightforward.
- The most attractive feature of finite differences is that it can be very easy to implement.
- There are several ways one could consider the FDM a special case of the FEM approach. E.g., first order FEM is identical to FDM for Poisson's equation, if the problem is discretised by a regular rectangular mesh with each rectangle divided into two triangles.
- There are reasons to consider the mathematical foundation of the finite element approximation more sound, for instance, because the quality of the approximation between grid points is poor in FDM.

- The quality of a FEM approximation is often higher than in the corresponding FDM approach, but this is extremely problem-dependent and several examples to the contrary can be provided.

Generally, FEM is the method of choice in all types of analysis in structural mechanics (i.e. solving for deformation and stresses in solid bodies or dynamics of structures) while computational fluid dynamics (CFD) tends to use FDM or other methods like finite volume method (FVM). CFD problems usually require discretisation of the problem into a large number of cells/gridpoints (millions and more), therefore cost of the solution favours simpler, lower order approximation within each cell. This is especially true for 'external flow' problems, like air flow around the car or airplane, or weather simulation.

Application

A variety of specialisations under the umbrella of the mechanical engineering discipline (such as aeronautical, biomechanical, and automotive industries) commonly use integrated FEM in design and development of their products. Several modern FEM packages include specific components such as thermal, electromagnetic, fluid, and structural working environments. In a structural simulation, FEM helps tremendously in producing stiffness and strength visualisations and also in minimising weight, materials, and costs.

FEM allows detailed visualisation of where structures bend or twist, and indicates the distribution of stresses and displacements. FEM software provides a wide range of simulation options for controlling the complexity of both modelling and analysis of a system. Similarly, the desired level of accuracy required and associated computational time requirements can be managed simultaneously to address most engineering applications. FEM allows entire designs to be constructed, refined, and optimised before the design is manufactured.

This powerful design tool has significantly improved both the standard of engineering designs and the methodology of the design process in many industrial applications. The introduction of FEM has substantially decreased the time to take products from concept to the production line. It is primarily through improved initial prototype designs using FEM that testing and development have been accelerated. In summary, benefits of FEM include increased accuracy, enhanced design and better insight into critical design parameters, virtual prototyping, fewer hardware prototypes, a faster and less expensive design cycle, increased productivity, and increased revenue. FEA has also been proposed to use in stochastic modelling, for numerically solving probability models.

Perturbation Theory

Perturbation theory comprises mathematical methods that are used to find an approximate solution to a problem which cannot be solved exactly, by starting from the exact solution of a related problem. Perturbation theory is applicable if the problem at hand can be formulated by adding a "small" term to the mathematical description of the exactly solvable problem.

Perturbation theory leads to an expression for the desired solution in terms of a formal power series in some "small" parameter – known as a perturbation series – that quantifies the deviation from the exactly solvable problem. The leading term in this power series is the solution of the exactly solvable problem, while further terms describe the deviation in the solution, due to the deviation from the initial problem. Formally, we have for the approximation to the full solution A, a series in the small parameter (here called ϵ), like the following:

$$A = A_0 + \epsilon^1 A_1 + \epsilon^2 A_2 + \cdots$$

In this example, A_0 would be the known solution to the exactly solvable initial problem and A_1, A_2... represent the higher-order terms which may be found iteratively by some systematic procedure. For small these higher-order terms in the series become successively smaller. An approximate "perturbation solution" is obtained by truncating the series, usually by keeping only the first two terms, the initial solution and the "first-order" perturbation correction:

$$A \approx A_0 + \epsilon A_1$$

General Description

Perturbation theory is closely related to methods used in numerical analysis. The earliest use of what would now be called *perturbation theory* was to deal with the otherwise unsolvable mathematical problems of celestial mechanics: Newton's solution for the orbit of the Moon, which moves noticeably differently from a simple Keplerian ellipse because of the competing gravitation of the Earth and the Sun.

Perturbation methods start with a simplified form of the original problem, which is *simple enough* to be solved exactly. In celestial mechanics, this is usually a Keplerian ellipse. Under non relativistic gravity, an ellipse is exactly correct when there are only two gravitating bodies (say, the Earth and the Moon) but not quite correct when there are three or more objects (say, the Earth, Moon, Sun, and the rest of the solar system).

The solved, but simplified problem is then *"perturbed"* to make the conditions that the perturbed solution actually satisfies closer to the real problem, such as including the gravitational attraction of a third body (the Sun). The "conditions" are a formula (or several) that represent reality, often something arising from a physical law like Newton's second law, the force-acceleration equation:

$$F = ma$$

In the case of the example, the force F is calculated based on the number of gravitationally relevant bodies; the acceleration a is obtained, using calculus, from the path of the Moon in its orbit. Both of these come in two forms: approximate values for force and acceleration, which result from simplifications, and hypothetical exact values for force and acceleration, which would require the complete answer to calculate.

The slight changes that result from accommodating the perturbation, which themselves may have been simplified yet again, are used as corrections to the approximate solution. Because of simplifications introduced along every step of the way, the corrections are never perfect, and the conditions met by the corrected solution do not perfectly match the equation demanded by reality, but even one cycle of corrections often provides a remarkably better approximate answer to what the real solution should be.

There is no requirement to stop at only one cycle of corrections. A partially corrected solution can be re-used as the new starting point for yet another cycle of perturbations and corrections. In principle, cycles of finding increasingly better corrections could go on indefinitely. In practice, one typically stops at one or two cycles of corrections. The usual difficulty with the method is that the corrections progressively make the new solutions very much more complicated, so each cycle is much more difficult to manage than the previous cycle of corrections. Isaac Newton is reported to have said, regarding the problem of the Moon's orbit, that *"It causeth my head to ache."*

This general procedure is a widely used mathematical tool in advanced sciences and engineering: start with a simplified problem and gradually add corrections that make the formula that the corrected problem matches closer and closer to the formula that represents reality. It is the natural extension to mathematical functions of the "guess, check, and fix" method used by older civilisations to compute certain numbers, such as square roots.

Examples

Examples for the "mathematical description" are: an algebraic equation, a differential equation (e.g., the equations of motion in celestial mechanics or a wave equation), a free energy (in statistical mechanics), a Hamiltonian operator (in quantum mechanics).

Examples for the kind of solution to be found perturbatively: the solution of the equation (e.g., the trajectory of a particle), the statistical average of some physical quantity (e.g., average magnetisation), the ground state energy of a quantum mechanical problem.

Examples for the exactly solvable problems to start with: linear equations, including linear equations of motion (harmonic oscillator, linear wave equation), statistical or quantum-mechanical systems of non-interacting particles (or in general, Hamiltonians or free energies containing only terms quadratic in all degrees of freedom).

Examples of "perturbations" to deal with: Nonlinear contributions to the equations of motion, interactions between particles, terms of higher powers in the Hamiltonian/Free Energy.

For physical problems involving interactions between particles, the terms of the perturbation series may be displayed (and manipulated) using Feynman diagrams.

History

Perturbation theory has its roots in early celestial mechanics, where the theory of epicycles was used to make small corrections to the predicted paths of planets. Curiously, it was the need for more and more epicycles that eventually led to the 16th century Copernican revolution in the understanding of planetary orbits. The development of basic perturbation theory for differential equations was fairly complete by the middle of the 19th century.

It was at that time that Charles-Eugène Delaunay was studying the perturbative expansion for the Earth-Moon-Sun system, and discovered the so-called "problem of small denominators". Here, the denominator appearing in the n'th term of the perturbative expansion could become arbitrarily small, causing the n'th correction to be as large or larger than the first-order correction. At the turn of the 20th century, this problem led Henri Poincaré to make one of the first deductions of the existence of chaos, or what is prosaically called the "butterfly effect": that even a very small perturbation can have a very large effect on a system.

Perturbation theory saw a particularly dramatic expansion and evolution with the arrival of quantum mechanics. Although perturbation theory was used in the semi-classical theory of the Bohr atom, the calculations were monstrously complicated, and subject to somewhat ambiguous interpretation. The discovery of Heisenberg's matrix mechanics allowed a vast simplification of the application of perturbation theory. Notable examples are the Stark effect and the Zeeman effect, which have a simple enough theory to be included in standard undergraduate textbooks in quantum mechanics. Other early applications include the fine structure and the hyperfine structure in the hydrogen atom.

In modern times, perturbation theory underlies much of quantum chemistry and quantum field theory. In chemistry, perturbation theory was used to obtain the first solutions for the helium atom.

In the middle of the 20th century, Richard Feynman realised that the perturbative expansion could be given a dramatic and beautiful graphical representation in terms of what are now called Feynman diagrams. Although originally applied only in quantum field theory, such diagrams now find increasing use in any area where perturbative expansions are studied.

A partial resolution of the small-divisor problem was given by the statement of the KAM theorem in 1954. Developed by Andrey Kolmogorov, Vladimir Arnold and Jürgen Moser, this theorem stated the conditions under which a system of partial differential equations will have only mildly chaotic behaviour under small perturbations.

In the late 20th century, broad dissatisfaction with perturbation theory in the quantum physics community, including not only the difficulty of going beyond second order in the expansion, but also questions about whether the perturbative expansion is even convergent, has led to a strong interest in the area of non-perturbative analysis, that is, the study of exactly solvable models. The prototypical model is the Korteweg–de Vries equation, a highly non-linear equation for which the interesting solutions, the solitons, cannot be reached by perturbation theory, even if the perturbations were carried out to infinite order. Much of the theoretical work in non-perturbative analysis goes under the name of quantum groups and non-commutative geometry.

Perturbation Orders

The standard exposition of perturbation theory is given in terms of the order to which the perturbation is carried out: first-order perturbation theory or second-order perturbation theory, and whether the perturbed states are degenerate (that is, singular), in which case extra care must be taken, and the theory is slightly more difficult.

This section needs to be expanded to include the standard textbook examples of each of the three expansions.

Perturbation Theory of Degenerate States

One may notice that the problem occurs in the first order perturbation theory when two or more eigenfunctions of the unperturbed system correspond to one eigenvalue i.e. when the eigenvalue equation becomes

$$D^{(0)} f_{n,i}^{(0)}(x) = \lambda_n^{(0)} f_{n,i}^{(0)}(x)$$

and the index i labels many states with the same eigenvalue $\lambda_n^{(0)}$. Expression for the eigenfunctions having the energy differences in the denominators becomes infinite. In that case the degenerate perturbation theory must be applied. The degeneracy must be removed first for higher order perturbation theory. The function is first assumed to be the linear combination of eigenfunctions with the same eigenvalue only

$$g(x) = \sum_k c_{n,k} f_{n,k}^{(0)}(x)$$

which again from the orthogonality of $f_{n,k}^{(0)}$ leads to the following equation

$$c_{n,i}\lambda_{n,i}^{(0)} + \epsilon \sum_k c_{n,k} \int f_{n,i}^{(0)}(x) D^{(1)} f_{n,k}^{(0)}(x)dx = \lambda c_{n,i}$$

for each n. As for the majority of low quantum numbers the changes over small range of integers the later equation can be usually solved analytically as at most 4×4 matrix equation. Once the degeneracy is removed the first and any order of the perturbation theory may be further used with respect to the new functions.

Commentary

Both regular and singular perturbation theory are frequently used in physics and engineering. Regular perturbation theory may only be used to find those solutions of a problem that evolve smoothly out of the initial solution when changing the parameter (that are "adiabatically connected" to the initial solution). A well known example from physics where regular perturbation theory fails is in fluid dynamics when one treats the viscosity as a small parameter. Close to a boundary, the fluid velocity goes to zero, even for very small viscosity (the no-slip condition).

For zero viscosity, it is not possible to impose this boundary condition and a regular perturbative expansion amounts to an expansion about an unrealistic physical solution. Singular perturbation theory can, however, be applied here and this amounts to 'zooming in' at the boundaries (using the method of matched asymptotic expansions).

Perturbation theory can fail when the system can transition to a different "phase" of matter, with a qualitatively different behaviour, that cannot be modelled by the physical formulas put into the perturbation theory (e.g., a solid crystal melting into a liquid).

In some cases, this failure manifests itself by divergent behaviour of the perturbation series. Such divergent series can sometimes be resummed using techniques such as Borel resummation. Perturbation techniques can be also used to find approximate solutions to non-linear differential equations. Examples of techniques used to find approximate solutions to these types of problems are the Lindstedt–Poincaré technique, the method of harmonic balancing, and the method of multiple time scales. There is absolutely no guarantee that perturbative methods result in a convergent solution. In fact, asymptotic series are the norm.

Perturbation Theory in Chemistry

Many of the ab initio quantum chemistry methods use perturbation theory directly or are closely related methods. Møller–Plesset perturbation theory uses the difference between the Hartree–Fock Hamiltonian and the exact non-relativistic Hamiltonian as the perturbation. The zero-order energy is the sum of orbital energies. The first-order energy is the Hartree–Fock energy and electron correlation is included at second-order or higher. Calculations to second, third or fourth order are very common and the code is included in most ab initio quantum chemistry programs. A related but more accurate method is the coupled cluster method.

Adomian Decomposition Method

The Adomian decomposition method (ADM) is a non-numerical method for solving nonlinear differential equations, both ordinary and partial. The general direction of this work is towards a unified theory for partial differential equations (PDE). The method was developed by George Adomian, chair of the Centre for Applied Mathematics at the University of Georgia. This method is a semi-analytical method.

Algebraic Differential Equation

In mathematics, an algebraic differential equation is a differential equation that can be expressed by means of differential algebra. There are several such notions, according to the concept of differential algebra used.

The intention is to include equations formed by means of differential operators, in which the coefficients are rational functions of the variables (e.g. the hypergeometric equation). Algebraic differential equations are widely used in computer algebra and number theory.

A simple concept is that of a polynomial vector field, in other words a vector field expressed with respect to a standard co-ordinate basis as the first partial derivatives with polynomial coefficients. This is a type of first-order algebraic differential operator.

Formulations

- Derivations D can be used as algebraic analogues of the formal part of differential calculus, so that algebraic differential equations make sense in commutative rings.
- The theory of differential fields was set up to express differential Galois theory in algebraic terms.
- The Weyl algebra W of differential operators with polynomial coefficients can be considered; certain modules M can be used to express differential equations, according to the presentation of M.
- The concept of Koszul connection is something that transcribes easily into algebraic geometry, giving an algebraic analogue of the way systems of differential equations are geometrically represented by vector bundles with connections.
- The concept of jet can be described in purely algebraic terms, as was done in part of Grothendieck's EGA project.
- The theory of D-modules is a global theory of linear differential equations, and has been developed to include substantive results in the algebraic theory (including a Riemann-Hilbert correspondence for higher dimensions).

Algebraic Solutions

It is usually not the case that the general solution of an algebraic differential equation is an algebraic function: solving equations typically produces novel transcendental functions. The case of algebraic solutions is however of considerable interest; the classical Schwarz list deals with the case of the hypergeometric equation.

In differential Galois theory the case of algebraic solutions is that in which the differential Galois group G is finite (equivalently, of dimension 0, or of a finite monodromy group for the case of Riemann surfaces and linear equations). This case stands in relation with the whole theory roughly as invariant theory does to group representation theory. The group G is in general difficult to compute, the understanding of algebraic solutions is an indication of upper bounds for G.

Autonomous System (Mathematics)

In mathematics, an autonomous system or autonomous differential equation is a system of ordinary differential equations which does not explicitly depend on the independent variable. When the variable is the time, they are also named Time-invariant system.

Many laws in physics, where the independent variable is usually assumed to be time, are expressed as autonomous systems because it is assumed the laws of nature which hold now are identical to those for any point in the past or future.

Autonomous systems are closely related to dynamical systems. Any autonomous system can be transformed into a dynamical system and, using very weak assumptions, a dynamical system can be transformed into an autonomous system.

Definition

An autonomous system is a system of ordinary differential equations of the form

$$\frac{d}{dt}x(t) = f(x(t))$$

where x takes values in n-dimensional Euclidean space and t is usually time.

It is distinguished from systems of differential equations of the form

$$\frac{d}{dt}x(t) = g(x(t), t)$$

in which the law governing the rate of motion of a particle depends not only on the particle's location, but also on time; such systems are not autonomous.

Qualitative Analysis

Autonomous systems can be analysed qualitatively using the phase space; in the one-variable case, this is the phase line.

Solution Techniques

The following techniques apply to one-dimensional autonomous differential equations. Any one-dimensional equation of order n is equivalent to an -dimensional first-order system (as described in Ordinary differential equation#Reduction to a first order system), but not necessarily vice versa.

First Order

The first-order autonomous equation

$$\frac{dx}{dt} = f(x)$$

is separable, so it can easily be solved by rearranging it into the integral form

$$t + C = \int \frac{dx}{f(x)}$$

Second Order

The second-order autonomous equation

$$\frac{d^2x}{dt^2} = f(x, x')$$

is more difficult, but it can be solved by introducing the new variable

$$v = \frac{dx}{dt}$$

and expressing the second derivative of x (via the chain rule) as

$$\frac{d^2x}{dt^2} = \frac{dv}{dt} = \frac{dx}{dt}\frac{dv}{dx} = v\frac{dv}{dx}$$

so that the original equation becomes

$$v\frac{dv}{dx} = f(x,v)$$

which is a first order equation containing no reference to the independent variable t and if solved provides u as a function of. Then, recalling the definition of:

$$\frac{dx}{dt} = v(x) \quad \Rightarrow \quad t + C = \int \frac{dx}{v(x)}$$

which is an implicit solution.

Special case: $x'' = f(x)$

The special case where f is independent of x'

$$\frac{d^2x}{dt^2} = f(x)$$

benefits from separate treatment. These types of equations are very common in classical mechanics because they are always Hamiltonian systems.

The idea is to make use of the identity (barring division by zero issues)

$$\frac{dx}{dt} = \left(\frac{dt}{dx}\right)^{-1}$$

which follows from the chain rule. Note aside then that by inverting both sides of a first order autonomous system, one can immediately integrate with respect to x:

$$\frac{dx}{dt} = f(x) \quad \Rightarrow \quad \frac{dt}{dx} = \frac{1}{f(x)} \quad \Rightarrow \quad t + C = \int \frac{dx}{f(x)}$$

which is another way to view the separation of variables technique. A natural question is then: can we do something like this with higher order equations?

The answer is yes for second order equations, but there's more work to do.

The second derivative must be expressed as a derivative with respect to x instead of t:

$$\frac{d^2x}{dt^2} = \frac{d}{dt}\left(\frac{dx}{dt}\right) = \frac{d}{dx}\left(\frac{dx}{dt}\right)\frac{dx}{dt} = \frac{d}{dx}\left(\left(\frac{dt}{dx}\right)^{-1}\right)\left(\frac{dt}{dx}\right)^{-1} =$$

$$= -\left(\frac{dt}{dx}\right)^{-2}\frac{d^2t}{dx^2}\left(\frac{dt}{dx}\right)^{-1} = -\left(\frac{dt}{dx}\right)^{-3}\frac{d^2t}{dx^2} =$$

$$= \frac{d}{dx}\left(\frac{1}{2}\left(\frac{dt}{dx}\right)^{-2}\right)$$

To reemphasize: what's been accomplished is that the second derivative in has been expressed as a derivative in.

The original second order equation may then finally be integrated:

$$\frac{d^2x}{dt^2} = f(x)$$

$$\frac{d}{dx}\left(\frac{1}{2}\left(\frac{dt}{dx}\right)^{-2}\right) = f(x)$$

$$\left(\frac{dt}{dx}\right)^{-2} = 2\int f(x)dx + C_1$$

$$\frac{dt}{dx} = \pm\frac{1}{\sqrt{2\int f(x)dx + C_1}}$$

$$t + C_2 = \pm\int\frac{dx}{\sqrt{2\int f(x)dx + C_1}}$$

This is an implicit solution, and beyond that the greatest potential problem is inability to simplify the integrals, which implies difficulty or impossibility in evaluating the integration constants.

Special Case: x'' = x'ⁿ f(x)

Using the above mentality, we can extend the technique to the more general equation

$$\frac{d^2x}{dt^2} = \left(\frac{dx}{dt}\right)^n f(x)$$

where n is some parameter not equal to two. This will work since the second derivative can be written in a form involving a power of x'.

Rewriting the second derivative, rearranging, and expressing the left side as a derivative:

$$-\left(\frac{dt}{dx}\right)^{-3}\frac{d^2t}{dx^2}=\left(\frac{dt}{dx}\right)^{-n}f(x)$$

$$-\left(\frac{dt}{dx}\right)^{n-3}\frac{d^2t}{dx^2}=f(x)$$

$$\frac{d}{dx}\left(\frac{1}{2-n}\left(\frac{dt}{dx}\right)^{n-2}\right)=f(x)$$

$$\left(\frac{dt}{dx}\right)^{n-2}=(2-n)\int f(x)dx+C_1$$

$$t+C_2=\int\left((2-n)\int f(x)dx+C_1\right)^{\frac{1}{n-2}}dx$$

The right will carry +/- if is even. The treatment must be different if:

$$-\left(\frac{dt}{dx}\right)^{-1}\frac{d^2t}{dx^2}=f(x)$$

$$-\frac{d}{dx}\left(\ln\left(\frac{dt}{dx}\right)\right)=f(x)$$

$$\frac{dt}{dx}=C_1e^{-\int f(x)dx}$$

$$t+C_2=C_1\int e^{-\int f(x)dx}dx$$

Chapter 6

Discrete Least Squares Meshless Method

The discrete least squares meshless (DLSM) method is a newly introduced meshless method based on the least squares concept. The method is based on the minimisation of a least squares functional defined as the weighted summation of the squared residual of the governing differential equation and its boundary conditions at nodal points used to discretise the domain and its boundaries. While most of the existing meshless methods need background cells for numerical integration, DLSM did not require numerical integration procedure due to the use of discrete least squares method to discretise the governing differential equation. A Moving least squares (MLS) approximation method is used to construct the shape function making the approach a fully least squares based approach.

Arzani and Afshar developed DLSM method in 2006 for the solution of Poisson equation. Firoozjaee and Afshar proposed collocated discrete least squares meshless (CDLSM) method to solve elliptic partial differential equations and studied the effect of the collocation points on the convergence and accuracy of the method. The method can be considered as an extension the earlier method of DLSM by introducing a set of collocation points for the calculation of least squares functional. CDLSM was later used by Naisipour et al. to solve elasticity problems on irregular distribution of nodal points. Afshar and Lashckarbolok used the CDLSM method for the adaptive simulation of hyperbolic problems. A simple posteriori error indicator based on the value of the least squares functional and a node moving strategy was used and tested on 1-D hyperbolic problems. Shobeyri and Afshar simulated free surface problems using DLSM method. The method was then extended for adaptive simulation of two-dimensional shocked hyperbolic problems by Afshar and Firoozjaee. Also, Afshar et al. formulated node moving and node enrichment adaptive refinement strategy for solution of elasticity problems.

Amani et al. proposed Mixed Discrete Least Squares Meshless (MDLSM) formulation for solution of planar elasticity problems. In this approach, the differential equations governing the planar elasticity problems are written in terms of the stresses and displacements which are approximated independently using the same shape functions. Since the resulting governing equations are of the first order, both the displacement and stress boundary conditions are of the Dirichlet-type which is easily incorporated via a penalty method. Because least squares based algorithm of MDLSM method, the proposed method does not need to be satisfied the Ladyzhenskaya–Babuška–Brezzi (LBB) condition.

Lanchester's Laws

Lanchester's laws are mathematical formulae for calculating the relative strengths of a predator/prey pair. This article is concerned with military forces. The Lanchester equations are differential equations describing the time dependence of two armies' strengths A and B as a function of time, with the function depending only on A and B.

In 1916, during the height of World War I, Frederick Lanchester devised a series of differential equations to demonstrate the power relationships between opposing forces. Among these are what is known as *Lanchester's Linear Law* (for ancient combat) and *Lanchester's Square Law* (for modern combat with long-range weapons such as firearms).

Lanchester's Linear Law

In ancient combat, between phalanxes of men with spears, say, one man could only ever fight exactly one other man at a time. If each man kills, and is killed by, exactly one other, then the number of men remaining at the end of the battle is simply the difference between the larger army and the smaller, assuming identical weapons. The linear law also applies to unaimed fire into an enemy-occupied area. The rate of attrition depends on the density of the available targets in the target area as well as the number of weapons firing. If two forces, occupying the same land area and using the same weapons, fire randomly into the same target area, they will both suffer the same rate and number of casualties, until the smaller force is eventually eliminated: the greater probability of any one shot hitting the larger force is balanced by the greater number of shots directed at the smaller force.

Lanchester's Square Law

Description

With firearms engaging each other directly with aimed fire from a distance, they can attack multiple targets and can receive fire from multiple directions. The rate of attrition now depends only on the number of weapons firing. Lanchester determined that the power of such a force is proportional not to the number of units it has, but to the square of the number of units. This is known as Lanchester's Square Law.

More precisely, the law specifies the casualties a firing force will inflict over a period of time, relative to those inflicted by the opposing force. In its basic form, the law is only useful to predict outcomes and casualties by attrition. It does not apply to whole armies, where tactical deployment means not all troops will be engaged all the time. It only works where each man (or ship, unit or whatever) can kill only one equivalent enemy at a time (so it does not apply to machine guns, artillery or—an extreme case—nuclear weapons). The law requires an assumption that casualties build up over time: it does not work in situations in which opposing troops kill each other instantly, either by firing simultaneously or by one side getting off the first shot and inflicting multiple casualties. Note that Lanchester's Square Law does not apply to technological force, only numerical force; so it takes an N-squared-fold increase in quality to make up for an N-fold increase in quantity.

Example Equations

Suppose that two armies, Red and Blue, are engaging each other in combat. Red is firing a continuous stream of bullets at Blue. Meanwhile, Blue is firing a continuous stream of bullets at Red. Let symbol A represent the number of soldiers in the Red force at the beginning of the battle. Each one has *offensive firepower* α, which is the number of enemy soldiers it can knock out of battle (e.g., kill or incapacitate) per unit time. Likewise, Blue has B soldiers, each with offensive firepower β.

Lanchester's square law calculates the number of soldiers lost on each side using the following pair of equations. Here, dA/dt represents the rate at which the number of Red soldiers is changing at a particular instant in time. A negative value indicates the loss of soldiers. Similarly, dB/dt represents the rate of change in the number of Blue soldiers.

$$dA/dt = -\beta B$$

$$dB/dt = -\alpha A$$

Relation to the Salvo Combat Model

Lanchester's equations are related to the more recent Salvo combat model equations, with two main differences. First, Lanchester's original equations form a continuous time model, whereas the basic salvo equations form a discrete time model. In a gun battle, bullets or shells are typically fired in large quantities. Each round has a relatively low chance of hitting its target, and does a relatively small amount of damage. Therefore Lanchester's equations model gunfire as a stream of firepower that continuously weakens the enemy force over time. By comparison, cruise missiles typically are fired in relatively small quantities. Each one has a high probability of hitting its target, and carries a relatively powerful warhead. Therefore it makes more sense to model them as a discrete pulse (or salvo) of firepower in a discrete time model.

Second, Lanchester's equations include only offensive firepower, whereas the salvo equations also include defensive firepower. Given their small size and large number, it is not practical to intercept bullets and shells in a gun battle. By comparison, cruise missiles can be intercepted (shot down) by surface-to-air missiles and anti-aircraft guns. So it is important to include such active defences in a missile combat model.

Lanchester's Law in Use

In modern warfare, to take into account that to some extent both linear and the square apply often an exponent of 1.5 is used.

Laplace Transform

The Laplace transform is a widely used integral transform with many applications in physics and engineering. Denoted $\mathcal{L}\{f(t)\}$, it is a linear operator of a function $f(t)$ with a real argument t (t e– 0) that transforms it to a function $F(s)$ with a complex argument s. This transformation is essentially bijective for the majority of practical uses; the respective pairs of $f(t)$ and $F(s)$ are matched in tables. The Laplace transform has the useful property that many relationships and operations over the originals $f(t)$ correspond to simpler relationships and operations over the images $F(s)$. It is named after Pierre-Simon Laplace, who introduced the transform in his work on probability theory.

The Laplace transform is related to the Fourier transform, but whereas the Fourier transform expresses a function or signal as a series of modes of vibration (frequencies), the Laplace transform resolves a function into its moments. Like the Fourier transform, the Laplace transform is used for solving differential and integral equations.

In physics and engineering it is used for analysis of linear time-invariant systems such as electrical circuits, harmonic oscillators, optical devices, and mechanical systems. In such analyses, the Laplace transform is often interpreted as a transformation from the *time-domain*, in which inputs and outputs are functions of time, to the *frequency-domain*, where the same inputs and outputs are functions of complex angular frequency, in radians per unit time. Given a simple mathematical or functional description of an input or output to a system, the Laplace transform provides an alternative functional description that often simplifies the process of analysing the behaviour of the system, or in synthesising a new system based on a set of specifications.

History

The Laplace transform is named after mathematician and astronomer Pierre-Simon Laplace, who used a similar transform (now called z transform) in his work on probability theory. The current widespread use of the transform came about soon after World War II although it had been used in the 19th century by Abel, Lerch, Heaviside and Bromwich. The older history of similar transforms is as follows. From 1744, Leonhard Euler investigated integrals of the form

$$z = \int X(x)e^{ax}\,dx \quad \text{and} \quad z = \int X(x)x^{A}\,dx$$

as solutions of differential equations but did not pursue the matter very far. Joseph Louis Lagrange was an admirer of Euler and, in his work on integrating probability density functions, investigated expressions of the form

$$\int X(x)e^{-ax}a^{x}\,dx,$$

which some modern historians have interpreted within modern Laplace transform theory.

These types of integrals seem first to have attracted Laplace's attention in 1782 where he was following in the spirit of Euler in using the integrals themselves as solutions of equations. However, in 1785, Laplace took the critical step forward when, rather than just looking for a solution in the form of an integral, he started to apply the transforms in the sense that was later to become popular. He used an integral of the form:

$$\int x^{s}\phi(x)dx,$$

akin to a Mellin transform, to transform the whole of a difference equation, in order to look for solutions of the transformed equation. He then went on to apply the Laplace transform in the same way and started to derive some of its properties, beginning to appreciate its potential power.

Laplace also recognised that Joseph Fourier's method of Fourier series for solving the diffusion equation could only apply to a limited region of space as the solutions were periodic. In 1809, Laplace applied his transform to find solutions that diffused indefinitely in space.

Formal Definition

The Laplace transform of a function $f(t)$, defined for all real numbers t e– 0, is the function $F(s)$, defined by:

$$F(s) = \mathcal{L}\{f(t)\} = \int_0^{\infty} e^{-st} f(t)dt.$$

The parameter s is a complex number:

$s = \sigma + i\omega$, with real numbers σ and ω.

The meaning of the integral depends on types of functions of interest. A necessary condition for existence of the integral is that f must be locally integrable on $[0, \infty)$. For locally integrable functions that decay at infinity or are of exponential type, the integral can be understood as a (proper) Lebesgue integral. However, for many applications it is necessary to regard it as a conditionally convergent improper integral at ". Still more generally, the integral can be understood in a weak sense, and this is dealt with below.

One can define the Laplace transform of a finite Borel measure μ by the Lebesgue integral

$$(\mathcal{L}\mu)(s) = \int_{[0,\infty)} e^{-st}\, d\mu(t).$$

An important special case is where μ is a probability measure or, even more specifically, the Dirac delta function. In operational calculus, the Laplace transform of a measure is often treated as though the measure came from a distribution function f. In that case, to avoid potential confusion, one often writes

$$(\mathcal{L}f)(s) = \int_{0^-}^{\infty} e^{-st} f(t)\,dt$$

where the lower limit of 0^- is shorthand notation for

$$\lim_{\varepsilon \downarrow 0} \int_{-\varepsilon}^{\infty}.$$

This limit emphasises that any point mass located at 0 is entirely captured by the Laplace transform. Although with the Lebesgue integral, it is not necessary to take such a limit, it does appear more naturally in connection with the Laplace–Stieltjes transform.

Probability Theory

In pure and applied probability, the Laplace transform is defined by means of an expectation value. If X is a random variable with probability density function f, then the Laplace transform of f is given by the expectation

$$(\mathcal{L}f)(s) = E\left[e^{-sX}\right].$$

By abuse of language, this is referred to as the Laplace transform of the random variable X itself. Replacing s by $-t$ gives the moment generating function of X. The Laplace transform has applications throughout probability theory, including first passage times of stochastic processes such as Markov chains, and renewal theory.

Bilateral Laplace Transform

When one says "the Laplace transform" without qualification, the unilateral or one-sided transform is normally intended. The Laplace transform can be alternatively defined as the *bilateral Laplace transform* or two-sided Laplace transform by extending the limits of integration to be the entire real axis. If that is done the common unilateral transform simply becomes a special case of the bilateral transform where the definition of the function being transformed is multiplied by the Heaviside step function.

The bilateral Laplace transform is defined as follows:

$$F(s) = \mathcal{L}\{f(t)\} = \int_{-\infty}^{\infty} e^{-st} f(t)dt.$$

Inverse Laplace Transform

The inverse Laplace transform is given by the following complex integral, which is known by various names (the Bromwich integral, the Fourier-Mellin integral, and Mellin's inverse formula):

$$f(t) = \mathcal{L}^{-1}\{F(s)\} = \frac{1}{2\pi i}\lim_{T\to\infty}\int_{\gamma-iT}^{\gamma+iT} e^{st} F(s)ds,$$

where γ is a real number so that the contour path of integration is in the *region of convergence* of $F(s)$. An alternative formula for the inverse Laplace transform is given by Post's inversion formula.

Region of Convergence

If f is a locally integrable function (or more generally a Borel measure locally of bounded variation), then the Laplace transform $F(s)$ of f converges provided that the limit

$$\lim_{R\to\infty}\int_0^R f(t)e^{-ts}\,dt$$

exists. The Laplace transform converges absolutely if the integral

$$\int_0^\infty |f(t)e^{-ts}|dt$$

exists (as a proper Lebesgue integral). The Laplace transform is usually understood as conditionally convergent, meaning that it converges in the former instead of the latter sense.

The set of values for which $F(s)$ converges absolutely is either of the form Re$\{s\} > a$ or else Re$\{s\} \geq a$, where a is an extended real constant, $-\infty \leq a \leq \infty$. (This follows from the dominated convergence theorem). The constant a is known as the abscissa of absolute convergence, and depends on the growth behaviour of $f(t)$. Analogously, the two-sided transform converges absolutely in a strip of the form $a <$ Re$\{s\} < b$, and possibly including the lines Re$\{s\} = a$ or Re$\{s\} = b$. The subset of values of s for which the Laplace transform converges absolutely is called the region of absolute convergence or the domain of absolute convergence. In the two-sided case, it is sometimes called the strip of absolute convergence. The Laplace transform is analytic in the region of absolute convergence.

Similarly, the set of values for which $F(s)$ converges (conditionally or absolutely) is known as the region of conditional convergence, or simply the region of convergence (ROC). If the Laplace transform converges (conditionally) at $s = s_0$, then it automatically converges for all s with Re$\{s\} >$ Re$\{s_0\}$. Therefore the region of convergence is a half-plane of the form Re$\{s\} > a$, possibly including some points of the boundary line Re$\{s\} = a$. In the region of convergence Re$\{s\} >$ Re$\{s_0\}$, the Laplace transform of f can be expressed by integrating by parts as the integral

$$F(s) = (s - s_0)\int_0^{\infty} e^{-(s-s_0)t}\beta(t)dt, \quad \beta(u) = \int_0^{u} e^{-s_0 t} f(t)dt.$$

That is, in the region of convergence $F(s)$ can effectively be expressed as the absolutely convergent Laplace transform of some other function. In particular, it is analytic. A variety of theorems, in the form of Paley–Wiener theorems, exist concerning the relationship between the decay properties of f and the properties of the Laplace transform within the region of convergence. In engineering applications, a function corresponding to a linear time-invariant (LTI) system is *stable* if every bounded input produces a bounded output. This is equivalent to the absolute convergence of the Laplace transform of the impulse response function in the region Re$\{s\}$ e– 0. As a result, LTI systems are stable provided the poles of the Laplace transform of the impulse response function have negative real part.

Relationship to other Transforms

Laplace–Stieltjes transform: The (unilateral) Laplace–Stieltjes transform of a function $g : \mathrm{R} \to \mathrm{R}$ is defined by the Lebesgue–Stieltjes integral

$$\{\mathcal{L}^* g\}(s) = \int_0^\infty e^{-st} dg(t).$$

The function g is assumed to be of bounded variation. If g is the antiderivative of f:

$$g(x) = \int_0^x f(t)dt$$

then the Laplace–Stieltjes transform of g and the Laplace transform of f coincide. In general, the Laplace–Stieltjes transform is the Laplace transform of the Stieltjes measure associated to g. So in practice, the only distinction between the two transforms is that the Laplace transform is thought of as operating on the density function of the measure, whereas the Laplace–Stieltjes transform is thought of as operating on its cumulative distribution function.

Fourier Transform

The continuous Fourier transform is equivalent to evaluating the bilateral Laplace transform with imaginary argument $s = i\omega$ or $s = 2\pi fi$:

$$\hat{f}(\omega) = \mathcal{F}\{f(t)\}$$
$$[1em] = \mathcal{L}\{f(t)\}|_{s=i\omega} = F(s)|_{s=i\omega}$$
$$[1em] = \int_{-\infty}^{\infty} e^{-i\omega t} f(t)\mathrm{d}t.$$

This expression excludes the scaling factor $1/\sqrt{2\pi}$, which is often included in definitions of the Fourier transform. This relationship between the Laplace and Fourier transforms is often used to determine the frequency spectrum of a signal or dynamical system. The above relation is valid as stated if and only if the region of convergence (ROC) of $F(s)$ contains the imaginary axis, $\sigma = 0$. For example, the function $f(t) = \cos(\omega_0 t)$ has a Laplace transform $F(s) = s/(s^2 + \omega_0^2)$ whose ROC is $\mathrm{Re}(s) > 0$. As $s = i\omega$ is a pole of $F(s)$, substituting $s = i\omega$ in $F(s)$ does not yield the Fourier transform of $f(t)u(t)$, which is proportional to the Dirac delta-function ä$(\omega\text{-}\omega_0)$.

However, a relation of the form

$$\lim_{\sigma \to 0^+} F(\sigma + i\omega) = \hat{f}(\omega)$$

holds under much weaker conditions. For instance, this holds for the above example provided that the limit is understood as a weak limit of measures. General conditions relating the limit of the Laplace transform of a function on the boundary to the Fourier transform take the form of Paley-Wiener theorems.

Mellin Transform

The Mellin transform and its inverse are related to the two-sided Laplace transform by a simple change of variables. If in the Mellin transform

$$G(s) = \mathcal{M}\{g(\theta)\} = \int_0^\infty \theta^s g(\theta) \frac{d\theta}{\theta}$$

we set $\theta = e^{-t}$ we get a two-sided Laplace transform.

Borel transform

The integral form of the Borel transform

$$F(s) = \int_0^\infty f(z) e^{-sz} \, dz$$

is a special case of the Laplace transform for f an entire function of exponential type, meaning that

$$| f(z) | \leq A e^{B|z|}$$

for some constants A and B. The generalised Borel transform allows a different weighting function to be used, rather than the exponential function, to transform functions not of exponential type. Nachbin's theorem gives necessary and sufficient conditions for the Borel transform to be well defined.

Fundamental Relationships

Since an ordinary Laplace transform can be written as a special case of a two-sided transform, and since the two-sided transform can be written as the sum of two one-sided transforms, the theory of the Laplace-, Fourier-, Mellin-, and *Z-transforms* are at bottom the same subject.

However, a different point of view and different characteristic problems are associated with each of these four major integral transforms.

s-Domain Equivalent Circuits and Impedances

The Laplace transform is often used in circuit analysis, and simple conversions to the s-Domain of circuit elements can be made. Circuit elements can be transformed into impedances, very similar to phasor impedances.

Here is a summary of equivalents:

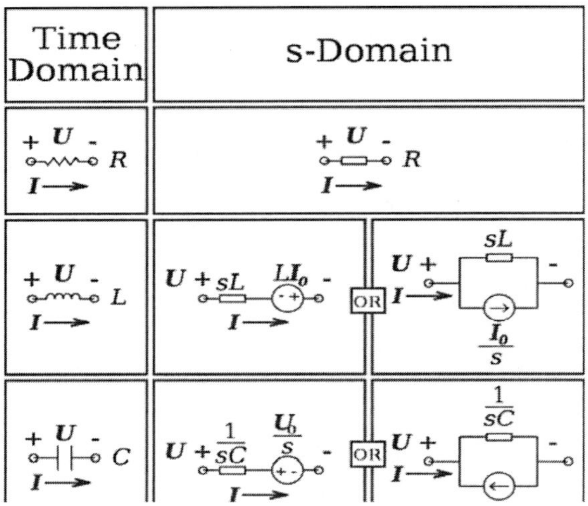

Note that the resistor is exactly the same in the time domain and the s-Domain. The sources are put in if there are initial conditions on the circuit elements.

For example, if a capacitor has an initial voltage across it, or if the inductor has an initial current through it, the sources inserted in the s-Domain account for that. The equivalents for current and voltage sources are simply derived from the transformations in the table above.

Examples: How to Apply the Properties and Theorems

The Laplace transform is used frequently in engineering and physics; the output of a linear time invariant system can be calculated by convolving its unit impulse response with the input signal. Performing this calculation in Laplace space turns the convolution into a multiplication; the latter being easier to solve because of its algebraic form. The Laplace transform can also be used to solve differential equations and is used extensively in electrical engineering.

The Laplace transform reduces a linear differential equation to an algebraic equation, which can then be solved by the formal rules of algebra. The original differential equation can then be solved by applying the inverse Laplace transform.

The English electrical engineer Oliver Heaviside first proposed a similar scheme, although without using the Laplace transform; and the resulting operational calculus is credited as the Heaviside calculus.

Example 1: Solving a Differential Equation

In nuclear physics, the following fundamental relationship governs radioactive decay:

the number of radioactive atoms N in a sample of a radioactive isotope decays at a rate proportional to N. This leads to the first order linear differential equation

$$\frac{dN}{dt} = -\lambda N$$

where λ is the decay constant. The Laplace transform can be used to solve this equation.

Rearranging the equation to one side, we have

$$\frac{dN}{dt} + \lambda N = 0.$$

Next, we take the Laplace transform of both sides of the equation:

$$\left(s\tilde{N}(s) - N_o\right) + \lambda\tilde{N}(s) = 0$$

where

$$\tilde{N}(s) = \mathcal{L}\{N(t)\}$$

and

$$N_o = N(0).$$

Solving, we find

$$\tilde{N}(s) = \frac{N_o}{s + \lambda}.$$

Finally, we take the inverse Laplace transform to find the general solution

$$N(t) = \mathcal{L}^{-1}\{\tilde{N}(s)\} = \mathcal{L}^{-1}\left\{\frac{N_o}{s+\lambda}\right\}$$
$$= N_o e^{-\lambda t},$$

which is indeed the correct form for radioactive decay.

Example 2: Deriving the Complex Impedance for a Capacitor

In the theory of electrical circuits, the current flow in a capacitor is proportional to the capacitance and rate of change in the electrical potential (in SI units). Symbolically, this is expressed by the differential equation

$$i = C\frac{dv}{dt}$$

where C is the capacitance (in farads) of the capacitor, $i = i(t)$ is the electric current (in amperes) through the capacitor as a function of time, and $v = v(t)$ is the voltage (in volts) across the terminals of the capacitor, also as a function of time. Taking the Laplace transform of this equation, we obtain

$$I(s) = C\left(sV(s) - V_o\right)$$

where

$$I(s) = \mathcal{L}\{i(t)\},$$
$$V(s) = \mathcal{L}\{v(t)\},$$

and

$$V_o = v(t)\,|_{t=0}\,.$$

Solving for $V(s)$ we have

$$V(s) = \frac{I(s)}{sC} + \frac{V_o}{s}.$$

The definition of the complex impedance Z (in ohms) is the ratio of the complex voltage V divided by the complex current I while holding the initial state V_o at zero:

$$Z(s) = \frac{V(s)}{I(s)}\Big|_{V_o=0}\,.$$

Using this definition and the previous equation, we find:

$$Z(s) = \frac{1}{sC},$$

which is the correct expression for the complex impedance of a capacitor.

Example 3: Method of Partial Fraction Expansion

Consider a linear time-invariant system with transfer function

$$H(s) = \frac{1}{(s+\alpha)(s+\beta)}.$$

The impulse response is simply the inverse Laplace transform of this transfer function:

$$h(t) = \mathcal{L}^{-1}\{H(s)\}.$$

To evaluate this inverse transform, we begin by expanding $H(s)$ using the method of partial fraction expansion:

$$\frac{1}{(s+\alpha)(s+\beta)} = \frac{P}{s+\alpha} + \frac{R}{s+\beta}.$$

The unknown constants P and R are the residues located at the corresponding poles of the transfer function. Each residue represents the relative contribution of that singularity to the transfer function's overall shape. By the residue theorem, the inverse Laplace transform depends only upon the poles and their residues. To find the residue P, we multiply both sides of the equation by $s + \alpha$ to get

$$\frac{1}{s+\beta} = P + \frac{R(s+\alpha)}{s+\beta}.$$

Then by letting $s = -\alpha$, the contribution from R vanishes and all that is left is

$$P = \left.\frac{1}{s+\beta}\right|_{s=-\alpha} = \frac{1}{\beta-\alpha}.$$

Similarly, the residue R is given by

$$R = \left.\frac{1}{s+\alpha}\right|_{s=-\beta} = \frac{1}{\alpha-\beta}.$$

Note that

$$R = \frac{-1}{\beta-\alpha} = -P$$

and so the substitution of R and P into the expanded expression for $H(s)$ gives

$$H(s) = \left(\frac{1}{\beta - \alpha}\right) \cdot \left(\frac{1}{s+\alpha} - \frac{1}{s+\beta}\right).$$

Finally, using the linearity property and the known transform for exponential decay, we can take the inverse Laplace transform of $H(s)$ to obtain:

$$h(t) = \mathcal{L}^{-1}\{H(s)\} = \frac{1}{\beta - \alpha}\left(e^{-\alpha t} - e^{-\beta t}\right),$$

which is the impulse response of the system.

Example 3.2: Convolution

The same result can be achieved using the convolution property as if the system is a series of filters with transfer functions of 1/(s+a) and 1/(s+b). That is, the inverse of

$$H(s) = \frac{1}{(s+a)(s+b)} = \frac{1}{s+a} \cdot \frac{1}{s+b}$$

is

$$\mathcal{L}^{-1}\{\frac{1}{s+a}\} * \mathcal{L}^{-1}\{\frac{1}{s+b}\} = e^{-at} * e^{-bt} = \int_0^t e^{-ax} e^{-b(t-x)}\, dx = \frac{e^{-at} - e^{-bt}}{b-a}.$$

Determining Structure of Astronomical Object from Spectrum

The wide and general applicability of the Laplace transform and its inverse is illustrated by an application in astronomy which provides some information on the *spatial distribution* of matter of an astronomical source of radiofrequency thermal radiation too distant to resolve as more than a point, given its flux density spectrum, rather than relating the *time* domain with the spectrum (frequency domain).

Assuming certain properties of the object, e.g. spherical shape and constant temperature, calculations based on carrying out an inverse Laplace transformation on the spectrum of the object can produce the only possible model of the distribution of matter in it (density as a function of distance from the centre) consistent with the spectrum. When independent information on the structure of an object is available, the inverse Laplace transform method has been found to be in good agreement.

Numerical Methods for Ordinary Differential Equations

Numerical ordinary differential equations is the part of numerical analysis which studies the numerical solution of ordinary differential equations (ODEs). This field is also known under the name *numerical integration,* but some people reserve this term for the computation of integrals.

Many differential equations cannot be solved analytically; however, in science and engineering, a numeric approximation to the solution is often good enough to solve a problem. The algorithms studied here can be used to compute such an approximation.

An alternative method is to use techniques from calculus to obtain a series expansion of the solution. Ordinary differential equations occur in many scientific disciplines, for instance in physics, chemistry, biology, and economics.

In addition, some methods in numerical partial differential equations convert the partial differential equation into an ordinary differential equation, which must then be solved.

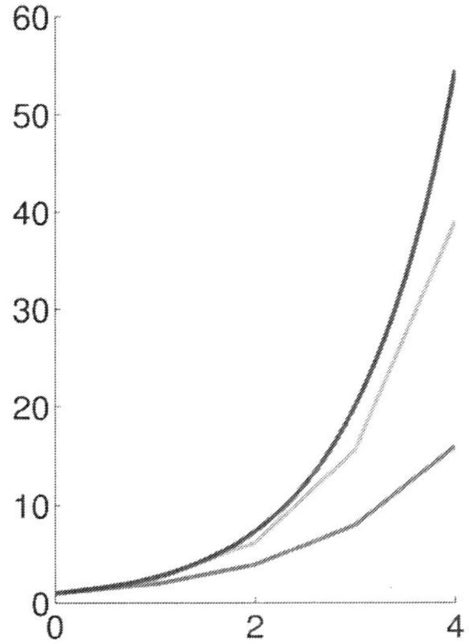

Figure: *Illustration of numerical integration for the differential equation* $y' = y, y(0) = 1$. *Blue: the Euler method, green: the midpoint method, red: the exact solution,* $y = e^t$. *The step size is* $h = 1.0$.

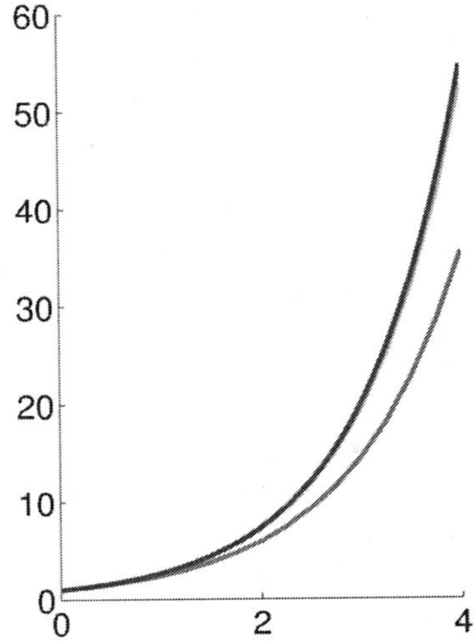

Figure: *The same illustration for* $h = 0.25$. *It is seen that the midpoint method converges faster than the Euler method.*

The Problem

We want to approximate the solution of the differential equation

$$y'(t) = f(t, y(t)), \qquad y(t_0) = y_0, \tag{1}$$

where f is a function that maps $[t_0,-) \times \mathrm{R}^d$ to R^d, and the initial condition $y_0 - \mathrm{R}^d$ is a given vector.

The above formulation is called an *initial value problem* (IVP). The Picard–Lindelöf theorem states that there is a unique solution, if f is Lipschitz continuous. In contrast, *boundary value problems* (BVPs) specify (components of) the solution y at more than one point. Different methods need to be used to solve BVPs, for example the shooting method (and its variants) or global methods like finite differences, Galerkin methods, or collocation methods.

Note that we restrict ourselves to *first-order* differential equations (meaning that only the first derivative of y appears in the equation, and no higher derivatives). This, however, does not restrict the generality of the problem, since a higher-order equation can easily be converted to a system of first-order equations by introducing extra variables. For example, the second-order equation $y'' = -y$ can be rewritten as two first-order equations: $y' = z$ and $z' = -y$.

Methods

Three elementary methods are discussed to give the reader a feeling for the subject. After that, pointers are provided to other methods (which are generally more accurate and efficient). The methods mentioned here are analysed in the next section.

The Euler Method

From any point on a curve, you can find an approximation of a nearby point on the curve by moving a short distance along a line tangent to the curve. Rigorous development: Starting with the differential equation (1), we replace the derivative y' by the finite difference approximation

$$y'(t) \approx \frac{y(t+h) - y(t)}{h}, \tag{2}$$

which when re-arranged yields the following formula

$$y(t+h) \approx y(t) + hy'(t)$$

and using (1) gives:

$$y(t+h) \approx y(t) + hf(t, y(t)). \tag{3}$$

This formula is usually applied in the following way. We choose a step size h, and we construct the sequence t_0, $t_1 = t_0 + h$, $t_2 = t_0 + 2h$, ... We denote by y_n a numerical estimate of the exact solution $y(t_n)$. Motivated by (3), we compute these estimates by the following recursive scheme

$$y_{n+1} = y_n + hf(t_n, y_n). \tag{4}$$

This is the *Euler method* (or *forward Euler method*, in contrast with the *backward Euler method*, to be described below). The method is named after Leonhard Euler who described it in 1768. The Euler method is an example of an *explicit* method. This means that the new value y_{n+1} is defined in terms of things that are already known, like y_n.

The Backward Euler Method

If, instead of (2), we use the approximation

$$y'(t) \approx \frac{y(t) - y(t-h)}{h}, \tag{5}$$

we get the *backward Euler method*:

$$y_{n+1} = y_n + hf(t_{n+1}, y_{n+1}). \tag{6}$$

The backward Euler method is an *implicit* method, meaning that we have to solve an equation to find y_{n+1}. One often uses fixed point iteration or (some modification of) the Newton–Raphson method to achieve this. Of course, it costs time to solve this equation; this cost must be taken into consideration when one selects the method to use. The advantage of implicit methods such as (6) is that they are usually more stable for solving a stiff equation, meaning that a larger step size h can be used.

The Exponential Euler Method

If the differential equation is of the form

$$y'(t) = A - By(t) \tag{7}$$

then an approximate explicit solution can be given by

$$y_{n+1} = y_n e^{-Bh} + \frac{A}{B}(1 - e^{-Bh}). \tag{8}$$

This method is commonly employed in neural simulations and it is the default integrator in the GENESIS neural simulator.

Generalisations

The Euler method is often not accurate enough. In more precise terms, it only has order one (the concept of *order* is explained below). This caused mathematicians to look for higher-order methods. One possibility is to use not only the previously computed value y_n to determine y_{n+1}, but to make the solution depend on more past values. This yields a so-called *multistep method*. Perhaps the simplest is the Leapfrog method which is second order and (roughly speaking) relies on two time values.

Almost all practical multistep methods fall within the family of linear multistep methods, which have the form

$$\alpha_k y_{n+k} + \alpha_{k-1} y_{n+k-1} + \cdots + \alpha_0 y_n$$
$$= h\left[\beta_k f(t_{n+k}, y_{n+k}) + \beta_{k-1} f(t_{n+k-1}, y_{n+k-1}) + \cdots + \beta_0 f(t_n, y_n)\right].$$

Another possibility is to use more points in the interval $[t_n, t_{n+1}]$. This leads to the family of Runge–Kutta methods, named after Carl Runge and Martin Kutta. One of their fourth-order methods is especially popular.

Advanced Features

A good implementation of one of these methods for solving an ODE entails more than the time-stepping formula.

It is often inefficient to use the same step size all the time, so *variable step-size methods* have been developed. Usually, the step size is chosen such that the (local) error per step is below some tolerance level. This means that the methods must also compute an *error indicator*, an estimate of the local error. An extension of this idea is to choose dynamically between different methods of different orders (this is called a *variable order method*). Methods based on Richardson extrapolation, such as the Bulirsch–Stoer algorithm, are often used to construct various methods of different orders.

Other desirable features include:

- *dense output*: cheap numerical approximations for the whole integration interval, and not only at the points t_0, t_1, t_2, ...
- *event location*: finding the times where, say, a particular function vanishes. This typically requires the use of a root-finding algorithm.
- support for parallel computing.
- when used for integrating with respect to time, time reversibility

Alternative Methods

Many methods do not fall within the framework discussed here. Some classes of alternative methods are:

- *multiderivative methods*, which use not only the function f but also its derivatives. This class includes *Hermite–Obreschkoff methods* and *Fehlberg methods*, as well as methods like the Parker–Sochacki method or Bychkov-Scherbakov method, which compute the coefficients of the Taylor series of the solution y recursively.
- *methods for second order ODEs*. We said that all higher-order ODEs can be transformed to first-order ODEs of the form (1). While this is certainly true, it may not be the best way to proceed. In particular, *Nyström methods* work directly with second-order equations.

- *geometric integration methods* are especially designed for special classes of ODEs (e.g., symplectic integrators for the solution of Hamiltonian equations). They take care that the numerical solution respects the underlying structure or geometry of these classes.

Analysis

Numerical analysis is not only the design of numerical methods, but also their analysis. Three central concepts in this analysis are:

- *convergence*: whether the method approximates the solution,
- *order*: how well it approximates the solution, and
- *stability*: whether errors are damped out.

Convergence

A numerical method is said to be *convergent* if the numerical solution approaches the exact solution as the step size h goes to 0. More precisely, we require that for every ODE (1) with a Lipschitz function f and every $t^* > 0$,

$$\lim_{h\to 0+} \max_{n=0,1,\ldots,\lfloor t^*/h \rfloor} \left\| y_{n,h} - y(t_n) \right\| = 0.$$

All the methods mentioned above are convergent. In fact, a numerical scheme has to be convergent to be of any use.

Consistency and Order

Suppose the numerical method is

$$y_{n+k} = \Psi(t_{n+k}; y_n, y_{n+1}, \ldots, y_{n+k-1}; h).$$

The *local (truncation) error* of the method is the error committed by one step of the method. That is, it is the difference between the result given by the method, assuming that no error was made in earlier steps, and the exact solution:

$$\delta^h_{n+k} = \Psi\left(t_{n+k}; y(t_n), y(t_{n+1}), \ldots, y(t_{n+k-1}); h\right) - y(t_{n+k}).$$

The method is said to be *consistent* if

$$\lim_{h\to 0} \frac{\delta^h_{n+k}}{h} = 0.$$

The method has *order* p if

$$\delta^h_{n+k} = O(h^{p+1}) \quad \text{as } h \to 0.$$

Hence a method is consistent if it has an order greater than 0. The (forward) Euler method (4) and the backward Euler method (6) introduced above both have order 1, so they are consistent. Most methods being used in practice attain higher order. Consistency is a necessary condition for convergence, but not sufficient; for a method to be convergent, it must be both consistent and zero-stable.

A related concept is the *global (truncation) error*, the error sustained in all the steps one needs to reach a fixed time t. Explicitly, the global error at time t is $y_N - y(t)$ where $N = (t-t_0)/h$. The global error of a pth order one-step method is $O(h^p)$; in particular, such a method is convergent. This statement is not necessarily true for multi-step methods.

Stability and Stiffness

For some differential equations, application of standard methods—such as the Euler method, explicit Runge–Kutta methods, or multistep methods (e.g., Adams–Bashforth methods)—exhibit instability in the solutions, though other methods may produce stable solutions. This "difficult behaviour" in the equation (which may not necessarily be complex itself) is described as *stiffness*, and is often caused by the presence of different time scales in the underlying problem. Stiff problems are ubiquitous in chemical kinetics, control theory, solid mechanics, weather forecasting, biology, plasma physics, and electronics.

History

Below is a timeline of some important developments in this field.

- 1768 - Leonhard Euler publishes his method.
- 1824 - Augustin Louis Cauchy proves convergence of the Euler method. In this proof, Cauchy uses the implicit Euler method.
- 1855 - First mention of the multistep methods of John Couch Adams in a letter written by F. Bashforth.
- 1895 - Carl Runge publishes the first Runge–Kutta method.
- 1905 - Martin Kutta describes the popular fourth-order Runge–Kutta method.

- 1910 - Lewis Fry Richardson announces his extrapolation method, Richardson extrapolation.
- 1952 - Charles F. Curtiss and Joseph Oakland Hirschfelder coin the term *stiff equations*.

Adaptive Mesh Refinement

In numerical analysis, adaptive mesh refinement is a method of adaptive meshing. Central to any Eulerian method is the manner in which it discretizes the continuous domain of interest into a grid of many individual elements. This grid may be *static*, established once and for all at the beginning of the computation, or it may be *dynamic*, tracking the features of the result as the computation progresses. If the computation has features which one wants to track which are much smaller than the overall scale of the problem, and which move in time, then one must either include many more static grids to cover the region of interest, or adopt a dynamic scheme.

The advantages of a dynamic gridding scheme are:

1. Increased computational savings over a static grid approach.
2. Increased storage savings over a static grid approach.
3. Complete control of grid resolution, compared to the fixed resolution of a static grid approach, or the Lagrangian-based adaptivity of smoothed particle hydrodynamics.

Introduction to Adaptive Mesh Refinement

In a series of papers, Marsha Berger, Joseph Oliger, and Phillip Colella developed an algorithm for dynamic gridding called *local adaptive mesh refinement*. The algorithm begins with the entire computational domain covered with a coarsely resolved base-level regular Cartesian grid. As the calculation progresses, individual grid cells are tagged for refinement, using a criterion that can either be user-supplied (for example mass per cell remains constant, hence higher density regions are more highly resolved) or based on Richardson extrapolation.

All tagged cells are then refined, meaning that a finer grid is overlaid on the coarse one. After refinement, individual grid patches on a single fixed level of refinement are passed off to an integrator which advances those cells in time. Finally, a correction procedure is implemented to correct the transfer along coarse-fine grid interfaces, to ensure that the amount of any conserved quantity leaving one cell exactly balances the amount entering the bordering cell. If at some point the level of refinement in a cell is greater than required, the high resolution grid may be removed and replaced with a coarser grid.

This allows the user to solve problems that are completely intractable on a uniform grid; for example, astrophysicists have used AMR to model a collapsing giant molecular cloud core down to an effective resolution of 131,072 cells per initial cloud radius, corresponding to a resolution of 10^{15} cells on a uniform grid.

Alternating Direction Implicit Method

In numerical analysis, the Alternating Direction Implicit (ADI) method is a finite difference method for solving parabolic and elliptic partial differential equations. It is most notably used to solve the problem of heat conduction or solving the diffusion equation in two or more dimensions. It is an example of an *operator splitting* method.

The traditional method for solving the heat conduction equation numerically is the Crank–Nicolson method. This method results in a very complicated set of equations in multiple dimensions, which are costly to solve. The advantage of the ADI method is that the equations that have to be solved in each step have a simpler structure and can be solved efficiently with the tridiagonal matrix algorithm.

The Method

Consider the linear diffusion equation in two dimensions,

$$\frac{\partial u}{\partial t} = \left(\frac{\partial^2 u}{\partial x^2} + \frac{\partial^2 u}{\partial y^2} \right) = (u_{xx} + u_{yy}) = \Delta u$$

The implicit Crank–Nicolson method produces the following finite difference equation:

$$\frac{u_{ij}^{n+1} - u_{ij}^{n}}{\Delta t} = \frac{1}{2}\left(\delta_x^2 + \delta_y^2\right)\left(u_{ij}^{n+1} + u_{ij}^{n}\right)$$

where δ_p is the central difference operator for the p-coordinate. After performing a stability analysis, it can be shown that this method will be stable for any Δt.

A disadvantage of the Crank–Nicolson method is that the matrix in the above equation is banded with a band width that is generally quite large. This makes direct solution of the system of linear equations quite costly (although efficient approximate solutions exist, for example use of the conjugate gradient method preconditioned with incomplete Cholesky factorisation).

The idea behind the ADI method is to split the finite difference equations into two, one with the x-derivative taken implicitly and the next with the y-derivative taken implicitly,

$$\frac{u_{ij}^{n+1/2} - u_{ij}^{n}}{\Delta t / 2} = \left(\delta_x^2 u_{ij}^{n+1/2} + \delta_y^2 u_{ij}^{n}\right)$$

$$\frac{u_{ij}^{n+1} - u_{ij}^{n+1/2}}{\Delta t / 2} = \left(\delta_x^2 u_{ij}^{n+1/2} + \delta_y^2 u_{ij}^{n+1}\right).$$

The system of equations involved is symmetric and tridiagonal (banded with bandwidth 3), and is typically solved using tridiagonal matrix algorithm.

It can be shown that this method is unconditionally stable and second order in time and space. There are more refined ADI methods such as the methods of Douglas, or the f-factor method which can be used for three or more dimensions.

Chapter 7

Boundary Knot Method

In numerical mathematics, the boundary knot method (BKM) is proposed as an alternative boundary-type meshfree distance function collocation scheme.

Recent decades have witnessed a research boom on the meshfree numerical PDE techniques since the construction of a mesh in the standard finite element method and boundary element method is not trivial especially for moving boundary, and higher-dimensional problems. The boundary knot method is different from the other methods based on the fundamental solutions, such as boundary element method, method of fundamental solutions and singular boundary method in that the former does not require special techniques to cure the singularity.

The BKM is truly meshfree, spectral convergent (numerical observations), symmetric (self-adjoint PDEs), integration-free, and easy to learn and implement. The method has successfully been tested to the Helmholtz, diffusion, convection-diffusion, and Possion equations with very irregular 2D and 3D domains.

Description

The BKM is basically a combination of the distance function, non-singular general solution, and dual reciprocity method (DRM). The distance function is employed in the BKM to approximate the inhomogeneous terms via the DRM, whereas the non-singular general solution of the partial differential equation leads to a boundary-only formulation for the homogeneous solution. Without the singular fundamental solution, the BKM removes the controversial artificial boundary in the method of fundamental solutions. Some preliminary numerical experiments show that the BKM can produce excellent results with relatively a small number of nodes for various linear and nonlinear problems.

History and Recent Developments

It has long been noted that boundary element method (BEM) is an alternative method to finite element method (FEM) and finite volume method (FVM) for infinite domain, thin-walled structures, and inverse problems, thanks to its dimensional reducibility. The major bottlenecks of BEM, however, are computationally expensive to evaluate integration of singular fundamental solution and to generate surface mesh or re-mesh. The method of fundamental solutions (MFS) has in recent decade emerged to alleviate these drawbacks and getting increasing attentions. The MFS is integration-free, spectral convergence and meshfree.

As its name implies, the fundamental solution of the governing equations is used as the basis function in the MFS. To avoid singularity of the fundamental solution, the artificial boundary outside the physical domain is required and has been a major bottleneck for the wide use of the MFS, since such fictitious boundary may cause computational instability. The BKM is classified as one kind of boundary-type meshfree methods without using mesh and artificial boundary.

The BKM has since been widely tested. In, the BKM is used to solve Laplace equation, Helmholtz Equation, and varying-parameter Helmholtz equations; in by analogy with Fasshauer's Hermite RBF interpolation, a symmetric BKM scheme is proposed in the presence of mixed boundary conditions; in, numerical investigations are made on the convergence of BKM in the analysis of homogeneous Helmholtz, modified Helmholtz and convection-diffusion problems; in the BKM is employed to deal with complicated geometry of two and three dimension Helmholtz and convection-diffusion problems; in membrane vibration under mixed-type boundary conditions is investigated by symmetric boundary knot method; in the BKM is applied to some inverse Helmholtz problems; in the BKM solves Poisson equations; in the BKM calculates Cauchy inverse inhomogeneous Helmholtz equations; in the BKM simulates the anisotropic problems via the geodesic distance; in relationships among condition number, effective condition number, and regularizations are investigated; in heat conduction in nonlinear functionally graded material is examined by the BKM; in the BKM is also used to solve nonlinear Eikonal equation.

Boundary Particle Method

In applied mathematics, the boundary particle method (BPM) is a truly boundary-only meshless (meshfree) collocation technique, in the sense that none of inner nodes are required at all in the numerical solution of nonhomogeneous partial differential equations. Numerical experiments also show that the BPM has spectral convergence. Its interpolation matrix can be symmetric and the method is easy-to-implement and free of integration and mesh. Thanks to its boundary-only merit, the BPM has clear edge over the other numerical schemes in the solution of optimisation and inverse problems, where only a part of boundary data is usually accessible.

History and Recent Developments

In recent decades, the dual reciprocity method (DRM) and multiple reciprocity method (MRM) have been emerging as the two most promising techniques to evaluate the particular solution of nonhomogeneous partial differential equations in conjunction with the boundary discretisation techniques, such as boundary element method (BEM). For instance, the so-called DR-BEM and MR-BEM are popular BEM techniques in the numerical solution of nonhomogeneous problems. The DRM has become de facto the method of choice in the boundary methods to evaluate the particular solution, since it is easy to use, efficient, and flexible to handle a variety of problems. However, the DRM demands the inner nodes to guarantee the convergence and stability. Therefore, the method is not truly boundary-only.

It has been claimed in literatures that the MRM has the striking advantage over the DRM in that it does not require using inner nodes at all for nonhomogeneous problems. However, the traditional MRM does also have disadvantages compared with the DRM. Firstly, the MRM is computationally much more expensive in the construction of the different interpolation matrices. Secondly, the method has limited applicability to general nonhomogeneous problems due to its conventional use of high-order Laplacian operators in the annihilation process.

An improved multiple reciprocity method, called the recursive composite multiple reciprocity method (RC-MRM), is proposed to overcome the above-mentioned problems. The key idea of the RC-MRM is employing high-order composite differential operators instead of high-order Laplacian operators to vanish a variety of nonhomogeneous terms in the governing equation, which can not otherwise be handled by the traditional MRM. In addition, the RC-MRM takes advantage of the recursive structures of the MRM interpolation matrix and significantly reduces computational costs.

The boundary particle method (BPM) is then developed to a boundary-only discretisation of inhomogeneous partial differential equation by combining the RC-MRM with a variety of the strong-form meshless boundary collocation discretisation schemes, such as the method of fundamental solution (MFS), boundary knot method (BKM), regularised meshless method (RMM), singular boundary method (SBM), and Trefftz method (TM). The BPM has since applied to a variety of problems such as nonhomogeneous Helmholtz and convection-diffusion equation. Numerical experiments are very encouraging. It is worthy of noting that the BPM interpolation representation is in fact of a wavelet series.

For the application of the BPM to Helmholtz, Poisson and plate bending problems, the high-order fundamental solution or general solution, harmonic function or Trefftz function (T-complete functions) are often used, for instance, those of Berger, Winkler, and vibrational thin plate equations. Thanks to its truly boundary-only merit, the BPM is more appealing in the solution of optimisation and inverse problems, where only a part of boundary data is usually accessible. The method has successfully been applied to inverse Cauchy problem associated with Poisson and nonhomogeneous Helmholtz equations.

Further Comments

The BPM may encounter difficulty in the solution of problems having complex source functions, such as non-smooth, large-gradient functions, or a set of discrete measured data. The road map for the BPM solution of such problems is briefly outlined below:

(1) The complex functions or a set of discrete measured data can be interpolated by a sum of polynomial or trigonometric function series. Then, the RC-MRM can easily reduce the nonhomogeneous equation to a high-order homogeneous equations, and the BPM can be simply implemented to solve these problems with boundary-only discretisation.

(2) The domain decomposition may be used to in the BPM boundary-only solution of large-gradient source functions problems.

G Space

G space is a functional space used in the formulation of general numerical methods based on meshfree methods and/ or finite element method settings. These numerical methods are applicable to solve PDEs in particular solid mechanics as well as fluid dynamics problems.

Description

For simplicity we choose elasticity problems for our discussion. In a weak formulation (such as in the FEM), displacement functions need to be in a proper Hilbert space, meaning that we need to make sure that the assumed displacement function is continuous over the entire problems domain. In a discrete setting (e.g., FEM), we construct the function using elements, but have to make sure that it is continuous along all element interfaces. This is also known as compatibility conditions. To ensure the compatibility, however, care must be taken, and FEM techniques should apply.

The G space theory accommodates functions that may be discontinuous. This is done by using the so-called generalised gradient smoothing technique, with which one can approximate the gradient of displacement functions in a proper G space. Since we do not have to actually perform even the 1st differentiation to the assumed displacement functions, the requirements on the consistence of the functions are further reduced, and hence the Weakened weak form or W2 form can be used to create stable and convergent computational methods. The stability is ensured by the so-called positivity conditions, and the convergence to the exact solution is ensure the admissible conditions on the assumed gradient (strain) fields.

History

The development of G Space theory started from the works on meshfree methods. The G Space theory forms the foundation for the W2 formulations, leading to various W2 models. The W2 models work well with triangular meshes and are insensitive to mesh distortion. Because triangular meshes can be generated automatically, the model becomes much easier in re-meshing and hence automation in modelling and simulation.

In addition, W2 models can be made soft enough (in uniform fashion) to produce upper bound solutions (for force-driving problems). Together with stiff models (such as the fully-compatible FEM models), one can conveniently bound the solution from both sides. This allows easy error estimation for generally complicated problems, as long as a triangular mesh can be generated.

Typical W2 models are the Smoothed Point Interpolation Methods (or S-PIM). The S-PIM can be node-based (known as NS-PIM or LC-PIM), edge-based (ES-PIM), and cell-based (CS-PIM). The NS-PIM was developed using the so-called SCNI technique. It was then discovered that NS-PIM is capable of producing upper bound solution and volumetric locking free. The ES-PIM is found superior in accuracy, and CS-PIM behaves in between the NS-PIM and ES-PIM. Moreover, W2 formulations allow the use of polynomial and radial basis functions in the creation of shape functions (it accommodates the discontinuous displacement functions, as long as it is in G1 space), which opens further room for future developments.

The S-FEM is largely the linear version of S-PIM, but with most of the properties of the S-PIM and much simpler. It also has variations of Node-based Smoothed FEM (NS-FEM) Edge-based Smoothed FEM (NS-FEM), Face-based Smoothed FEM (NS-FEM), Cell-based Smoothed FEM (NS-FEM), Edge/node-based Smoothed FEM (NS/ES-FEM), as well as Alpha FEM method (Alpha FEM).

Applications

Numerical methods built on G space theory have been applied to solve the following physical problems:

1) Mechanics for solids, structures and piezoelectrics;
2) Fracture mechanics and crack propagation;
3) Heat transfer;
4) Structural acoustics;
5) Nonlinear and contact problems;
6) Adaptive Analysis;
7) Phase change problem;
8) Limited analysis.

Meshfree Methods

Meshfree methods are a particular class of numerical simulation algorithms for the simulation of physical phenomena. Traditional simulation algorithms relied on a grid or a mesh, meshfree methods in contrast use the geometry of the simulated object directly for calculations. Meshfree methods exist for fluid dynamics as well as for solid mechanics. Some methods are able to handle both cases.

Description

Meshfree methods eliminate some or all of the traditional mesh-based view of the computational domain and rely on a particle (either Lagrangian or Eulerian) view of the field problem. A goal of meshfree methods is to facilitate the simulation of increasingly demanding problems that require the ability to treat large deformations, advanced materials, complex geometry, nonlinear material behaviour, discontinuities and singularities. For example the melting of a solid or the freezing process can be simulated using meshfree methods.

History and Recent Development

One of the earlier methods without a mesh is smoothed particle hydrodynamics, presented in 1977. Many methods listed in the next section are developed during the past 30 some years. Recent advances on meshfree methods aim at the development of computational tools for automation in modelling and simulations. This is enabled by the so-called weakened weak (W2) formulation based on the G space theory. The W2 formulation offers possibilities for formulate various (uniformly) "soft" models that works well with triangular meshes. Because triangular mesh can be generated automatically, it becomes much easier in re-meshing and hence automation in modelling and simulation. In addition, W2 models can be made soft enough (in uniform fashion) to produce upper bound solutions (for force-driving problems).

Together with stiff models (such as the fully compatible FEM models), one can conveniently bound the solution from both sides. This allows easy error estimation for generally complicated problems, as long as a triangular mesh can be generated. Typical W2 models are the Smoothed Point Interpolation Methods (or S-PIM). The S-PIM can be node-based (known as NS-PIM or LC-PIM), edge-based (ES-PIM), and cell-based (CS-PIM). The NS-PIM was developed using the so-called SCNI technique. It was then discovered that NS-PIM is capable of producing upper bound solution and volumetric locking free. The ES-PIM is found superior in accuracy, and CS-PIM behaves in between the NS-PIM and ES-PIM. Moreover, W2 formulations allow the use of polynomial and radial basis functions in the creation of shape functions (it accommodates the discontinuous displacement functions, as long as it is in G1 space), which opens further rooms for future developments.

The W2 formulation has also led to the development of combination of meshfree techniques with the well-developed FEM techniques, and one can now use triangular mesh with excellent accuracy and desired softness. A typical such a formulation is the so-called Smoothed Finite Element Method (or S-FEM) The S-FEM is the linear version of S-PIM, but with most of the properties of the S-PIM and much simpler.

It is a general perception that meshfree methods are much more expensive than the FEM counterparts. The recent study has found however, the S-PIM and S-FEM can be much faster than the FEM counterparts. The S-PIM and S-FEM works well for solid mechanics problems. For [CFD] problems, the formulation can be simpler, via strong formulation. A Gradient Smoothing Methods (GSM) has also be developed recently for [CFD] problems, implementing the gradient smoothing idea in strong form.

The GSM is similar to [FVM], but uses gradient smoothing operations exclusively in nested fashions, and is a general numerical method for PDEs.

Methods and Acronyms

The following numerical methods are generally considered to fall within the general class of "meshfree" methods. Acronyms are provided in parentheses.

- Smoothed particle hydrodynamics (SPH) (1977)
- Diffuse element method (DEM) (1992)
- Dissipative particle dynamics (DPD) (1992)
- Element-free Galerkin method (EFG / EFGM) (1994)
- Reproducing kernel particle method (RKPM) (1995)
- Finite pointset method (FPM) (1998)
- hp-clouds
- Natural element method (NEM)
- Material Point Method (MPM)
- Meshless local Petrov Galerkin (MLPG)
- Moving particle semi-implicit (MPS)
- Generalised finite difference method (GFDM)
- Particle-in-cell (PIC)
- Moving particle finite element method (MPFEM)
- Finite cloud method (FCM)
- Boundary node method (BNM)
- Meshfree moving Kriging interpolation method (MK)
- Boundary cloud method (BCM)
- Method of fundamental solution(MFS)
- Method of particular solution (MPS)
- Method of Finite Spheres (MFS)
- Discrete Vortex Method (DVM)
- Smoothed point interpolation method (S-PIM) (2005).
- Meshfree local radial point interpolation method (RPIM).
- Local Radial Basis Function Collocation Method (LRBFCM)
- Viscous vortex domains method (VVD)
- Discrete Least Squares Meshless method (DLSM) (2006)

Related methods:

- Moving least squares (MLS) – provide general approximation method for arbitrary set of nodes
- Partition of unity methods (PoUM) – provide general approximation formulation used in some meshfree methods
- Continuous blending method (enrichment and coupling of finite elements and meshless methods)
- eXtended FEM, Generalised FEM (XFEM, GFEM) – variants of FEM (finite element method) combining some meshless aspects
- Smoothed finite element method (S-FEM) (2007)
- Gradient smoothing method (GSM) (2008)
- Local maximum-entropy (LME)
- Space-Time Meshfree Collocation Method (STMCM)

Smoothed Finite Element Method

Smoothed Finite Element methods (S-FEM) are a particular class of numerical simulation algorithms for the simulation of physical phenomena. It was developed by combining meshfree methods with the finite element method. S-FEM are applicable to solid mechanics as well as fluid dynamics, but so far mainly applied for solid mechanics problems.

Description

The essential idea in the S-FEM is to use a finite element mesh (in particular triangular mesh) to construct numerical models of good performance. This is achieved by modifying the compatible strain field, or construct a strain field using only the displacements, hoping a Galerkin model using the modified/constructed strain field can deliver some good properties.

Such a modification/construction can be performed within elements but more often beyond the elements (meshfree concepts): bring in the information from the neighbouring elements.

Naturally, the strain field has to satisfy certain conditions, and the standard Galerkin weakform needs to be modified accordingly to ensure the stability and convergence.

History

The development of S-FEM started from the works on meshfree methods, where the so-called weakened weak (W2) formulation based on the G space theory were developed. The W2 formulation offers possibilities for formulate various (uniformly) "soft" models that works well with triangular meshes.

Because triangular mesh can be generated automatically, it becomes much easier in re-meshing and hence automation in modelling and simulation.

In addition, W2 models can be made soft enough (in uniform fashion) to produce upper bound solutions (for force-driving problems). Together with stiff models (such as the fully-compatible FEM models), one can conveniently bound the solution from both sides. This allows easy error estimation for generally complicated problems, as long as a triangular mesh can be generated. Typical W2 models are the Smoothed Point Interpolation Methods (or S-PIM).

The S-PIM can be node-based (known as NS-PIM or LC-PIM), edge-based (ES-PIM), and cell-based (CS-PIM). The NS-PIM was developed using the so-called SCNI technique. It was then discovered that NS-PIM is capable of producing upper bound solution and volumetric locking free.

The ES-PIM is found superior in accuracy, and CS-PIM behaves in between the NS-PIM and ES-PIM. Moreover, W2 formulations allow the use of polynomial and radial basis functions in the creation of shape functions (it accommodates the discontinuous displacement functions, as long as it is in G1 space), which opens further rooms for future developments.

The S-FEM is largely the linear version of S-PIM, but with most of the properties of the S-PIM and much simpler. It has also variations of NS-FEM, ES-FEM and CS-FEM. The major property of S-PIM can be found also in S-FEM.

S-FEM Models

- Node-based Smoothed FEM (NS-FEM)
- Edge-based Smoothed FEM (NS-FEM)
- Face-based Smoothed FEM (NS-FEM)
- Cell-based Smoothed FEM (NS-FEM)

- Node/Edge-based Smoothed FEM (NS/ES-FEM)
- Alpha FEM method (Alpha FEM)

Applications

S-FEM has been applied to solve the following physical problems:

1) Mechanics for solids, structures and piezoelectrics;
2) Fracture mechanics and crack propagation;
3) Heat transfer;
4) Structural acoustics;
5) Nonlinear and contact problems;
6) Adaptive Analysis;
7) Phase change problem;
8) Limited analysis.

Weakened Weak Form

Weakened weak form (or W2 form) is used in the formulation of general numerical methods based on meshfree methods and/or finite element method settings. These numerical methods are applicable to solid mechanics as well as fluid dynamics problems.

Description: For simplicity we choose elasticity problems (2nd order PDE) for our discussion. Our discussion is also most convenient in reference to the well-known weak and strong form. In a strong formulation for an approximate solution, we need to assume displacement functions that are 2nd order differentiable.

In a weak formulation, we create linear and bilinear forms and then search for a particular function (an approximate solution) that satisfy the weak statement. The bilinear form uses gradient of the functions that has only 1st order differentiation.

Therefore, the requirement on the continuity of assumed displacement functions is weaker than in the strong formulation. In a discrete form (such as the Finite element method, or FEM), a sufficient requirement for a assumed displacement function is piecewise continuous over the entire problems domain. This allows us to construct the function using elements (but making sure it is continuous a long all element interfaces), leading to the powerful FEM.

Now, in a weakened weak (W2) formulation, we further reduce the requirement. We form a bilinear form using only the assumed function (not even the gradient).

This is done by using the so-called generalised gradient smoothing technique, with which one can approximate the gradient of displacement functions for certain class of discontinuous functions, as long as they are in a proper G space.

Since we do not have to actually perform even the 1st differentiation to the assumed displacement functions, the requirement on the consistence of the functions are further reduced, and hence the weakened weak or W2 formulation.

History: The development of systematic theory of the weakened weak form started from the works on meshfree methods. It is relatively new, but had very rapid development in the past few years.

Features of W2 Formulations:

1) The W2 formulation offers possibilities for formulate various (uniformly) "soft" models that works well with triangular meshes. Because triangular mesh can be generated automatically, it becomes much easier in re-meshing and hence automation in modelling and simulation. This is very important for our long term goal of development of fully automated computational methods.
2) In addition, W2 models can be made soft enough (in uniform fashion) to produce upper bound solutions (for force-driving problems). Together with stiff models (such as the fully-compatible FEM models), one can conveniently bound the solution from both sides. This allows easy error estimation for generally complicated problems, as long as a triangular mesh can be generated. This is important for producing so-called certified solutions.
3) W2 models can be built free from volumetric locking, and possibly free from other types of locking phenomena.

4) W2 models provide the freedom to assume separately the displacement gradient of the displacement functions, offering opportunities for ultra-accurate and super-convergent models. It may be possible to construct linear models with energy convergence rate of 2.

5) W2 models are often found less sensitive to mesh distortion.

6) W2 models are found effective for low order methods.

Existing W2 Models: Typical W2 models are the Smoothed Point Interpolation Methods (or S-PIM). The S-PIM can be node-based (known as NS-PIM or LC-PIM), edge-based (ES-PIM), and cell-based (CS-PIM).

The NS-PIM was developed using the so-called SCNI technique. It was then discovered that NS-PIM is capable of producing upper bound solution and volumetric locking free.

The ES-PIM is found superior in accuracy, and CS-PIM behaves in between the NS-PIM and ES-PIM. Moreover, W2 formulations allow the use of polynomial and radial basis functions in the creation of shape functions (it accommodates the discontinuous displacement functions, as long as it is in G1 space), which opens further rooms for future developments.

The S-FEM is largely the linear version of S-PIM, but with most of the properties of the S-PIM and much simpler. It has also variations of NS-FEM, ES-FEM and CS-FEM.

The major property of S-PIM can be found also in S-FEM. The S-FEM models are:

- Node-based Smoothed FEM (NS-FEM)
- Edge-based Smoothed FEM (NS-FEM)
- Face-based Smoothed FEM (NS-FEM)
- Cell-based Smoothed FEM (NS-FEM)
- Edge/node-based Smoothed FEM (NS/ES-FEM)
- Alpha FEM method (Alpha FEM)

Applications: Some of the applications of W2 models are:

1) Mechanics for solids, structures and piezoelectrics </ ref> Liu GR, Nguyen-Xuan H, Nguyen-Thoi T, A theoretical study on the smoothed FEM (S-FEM) models:

Properties, accuracy and convergence rates, INTERNATIONAL JOURNAL FOR NUMERICAL METHODS IN ENGINEERING Vol. 84 Issue: 10, 1222-1256, 2010 </ref>;

2) Fracture mechanics and crack propagation;
3) Heat transfer;
4) Structural acoustics;
5) Nonlinear and contact problems;
6) Adaptive Analysis;
7) Phase change problem;
8) Limited analysis.

Chapter 8

Stochastic Differential Equation

A stochastic differential equation (SDE) is a differential equation in which one or more of the terms is a stochastic process, resulting in a solution which is itself a stochastic process. SDE are used to model diverse phenomena such as fluctuating stock prices or physical system subject to thermal fluctuations. Typically, SDEs incorporate white noise which can be thought of as the derivative of Brownian motion (or the Wiener process); however, it should be mentioned that other types of random fluctuations are possible, such as jump processes.

Background

The earliest work on SDEs was done to describe Brownian motion in Einstein's famous paper, and at the same time by Smoluchowski. However, one of the earlier works related to Brownian motion is credited to Bachelier (1900) in his thesis 'Theory of Speculation'. This work was followed upon by Langevin. Later Itô and Stratonovich put SDEs on more solid mathematical footing.

Terminology

In physical science, SDEs are usually written as Langevin equations. These are sometimes confusingly called "the Langevin equation" even though there are many possible forms. These consist of an ordinary differential equation containing a deterministic part and an additional random white noise term. A second form is the Smoluchowski equation and, more generally, the Fokker-Planck equation. These are partial differential equations that describe the time evolution of probability distribution functions.

The third form is the stochastic differential equation that is used most frequently in mathematics and quantitative finance. This is similar to the Langevin form, but it is usually written in differential form. SDEs come in two varieties, corresponding to two versions of stochastic calculus.

Stochastic Calculus

Brownian motion or the Wiener process was discovered to be exceptionally complex mathematically. The Wiener process is nowhere differentiable; thus, it requires its own rules of calculus. There are two dominating versions of stochastic calculus, the Ito stochastic calculus and the Stratonovich stochastic calculus. Each of the two has advantages and disadvantages, and newcomers are often confused whether the one is more appropriate than the other in a given situation. Guidelines exist (e.g. Ø ksendal, 2003) and conveniently, one can readily convert an Ito SDE to an equivalent Stratonovich SDE and back again. Still, one must be careful which calculus to use when the SDE is initially written down.

Numerical Solutions

Numerical solution of stochastic differential equations and especially stochastic partial differential equations is a young field relatively speaking. Almost all algorithms that are used for the solution of ordinary differential equations will work very poorly for SDEs, having very poor numerical convergence.

Methods include the Euler–Maruyama method, Milstein method and Runge–Kutta method (SDE).

Use in Physics

In physics, SDEs are typically written in the Langevin form and referred to as "the Langevin equation." For example, a general coupled set of first-order SDEs is often written in the form:

$$\dot{x}_i = \frac{dx_i}{dt} = f_i(\mathbf{x}) + \sum_{m=1}^{n} g_i^m(\mathbf{x})\eta_m(t),$$

where $\mathbf{x} = \{x_i \mid 1 \le i \le k\}$ is the set of unknowns, the f_i and g_i are arbitrary functions and the η_m are random functions of time, often referred to as "noise terms". This form is usually usable because there are standard techniques for transforming higher-order equations into several coupled first-order equations by introducing new unknowns. If the are constants, the system is said to be subject to additive noise, otherwise it is said to be subject to multiplicative noise. This term is somewhat misleading as it has come to mean the general case even though it appears to imply the limited case where : $g(x) \propto x$. Additive noise is the simpler of the two cases; in that situation the correct solution can often be found using ordinary calculus and in particular the ordinary chain rule of calculus. However, in the case of multiplicative noise, the Langevin equation is not a well-defined entity on its own, and it must be specified whether the Langevin equation should be interpreted as an Ito SDE or a Stratonovich SDE.

In physics, the main method of solution is to find the probability distribution function as a function of time using the equivalent Fokker-Planck equation (FPE). The Fokker-Planck equation is a deterministic partial differential equation. It tells how the probability distribution function evolves in time similarly to how the Schrödinger equation gives the time evolution of the quantum wave function or the diffusion equation gives the time evolution of chemical concentration. Alternatively numerical solutions can be obtained by Monte Carlo simulation. Other techniques include the path integration that draws on the analogy between statistical physics and quantum mechanics (for example, the Fokker-Planck equation can be transformed into the Schrödinger equation by rescaling a few variables) or by writing down ordinary differential equations for the statistical moments of the probability distribution function.

Note on "the Langevin Equation"

The "the" in "the Langevin equation" is somewhat ungrammatical nomenclature. Each individual physical model has its own Langevin equation. Perhaps, "a Langevin equation" or "the associated Langevin equation" would conform better with common English usage.

Use in Probability and Mathematical Finance

The notation used in probability theory (and in many applications of probability theory, for instance mathematical finance) is slightly different. This notation makes the exotic nature of the random function of time η_m in the physics formulation more explicit. It is also the notation used in publications on numerical methods for solving stochastic differential equations. In strict mathematical terms, can not be chosen as a usual function, but only as a generalised function. The mathematical formulation treats this complication with less ambiguity than the physics formulation.

A typical equation is of the form

$$dX_t = \mu(X_t, t)dt + \sigma(X_t, t)dB_t,$$

where denotes a Wiener process (Standard Brownian motion). This equation should be interpreted as an informal way of expressing the corresponding integral equation

$$X_{t+s} - X_t = \int_t^{t+s} \mu(X_u, u)du + \int_t^{t+s} \sigma(X_u, u)dB_u.$$

The equation above characterises the behaviour of the continuous time stochastic process X_t as the sum of an ordinary Lebesgue integral and an Itô integral. A heuristic (but very helpful) interpretation of the stochastic differential equation is that in a small time interval of length δ the stochastic process X_t changes its value by an amount that is normally distributed with expectation $\mu(X_t, t)\ \delta$ and variance $\sigma(X_t, t)^2\ \delta$ and is independent of the past behaviour of the process. This is so because the increments of a Wiener process are independent and normally distributed. The function μ is referred to as the drift coefficient, while σ is called the diffusion coefficient. The stochastic process X_t is called a diffusion process, and is usually a Markov process.

The formal interpretation of an SDE is given in terms of what constitutes a solution to the SDE. There are two main definitions of a solution to an SDE, a strong solution and a weak solution. Both require the existence of a process X_t that solves the integral equation version of the SDE. The difference between the two lies in the underlying probability space (Ω F, Pr). A weak solution consists of a probability space and a process that satisfies the integral equation, while a strong solution is a process that satisfies the equation and is defined on a given probability space. An important example is the equation for geometric Brownian motion

$$dX_t = \mu X_t\,dt + \sigma X_t\,dB_t.$$

which is the equation for the dynamics of the price of a stock in the Black Scholes options pricing model of financial mathematics.There are also more general stochastic differential equations where the coefficients μ and σ depend not only on the present value of the process X_t, but also on previous values of the process and possibly on present or previous values of other processes too. In that case the solution process, *X*, is not a Markov process, and it is called an Itô process and not a diffusion process. When the coefficients depends only on present and past values of *X*, the defining equation is called a stochastic delay differential equation.

Delay Differential Equation

In mathematics, delay differential equations (DDEs) are a type of differential equation in which the derivative of the unknown function at a certain time is given in terms of the values of the function at previous times.

A general form of the time-delay differential equation for $x(t) \in \mathbb{R}^n$ is

$$\frac{d}{dt}x(t) = f(t, x(t), x_t),$$

where $x_t = \{x(\tau) : \tau \le t\}$ represents the trajectory of the solution in the past. In this equation, f is a functional operator from $\mathbb{R} \times \mathbb{R}^n \times C^1$ to $\mathbb{R}^n$.

Examples

Continuous delay

$$\frac{d}{dt}x(t) = f\left(t, x(t), \int_{-\infty}^{0} x(t+\tau)\,d\mu(\tau)\right)$$

Discrete delay

$$\frac{d}{dt}x(t) = f(t, x(t), x(t-\tau_1), \ldots., x(t-\tau_m))$$

for $\tau_1 > \cdots > \tau_m \ge 0$.

Linear with discrete delay

$$\frac{\mathrm{d}}{\mathrm{d}t}x(t) = A_0 x(t) + A_1 x(t-\tau_1) + \cdots + A_m x(t-\tau_m)$$

where $A_0, \ldots, A_m \in \mathbb{R}^{n\times n}$.

Pantograph equation

$$\frac{\mathrm{d}}{\mathrm{d}t}x(t) = ax(t) + bx(\lambda t),$$

where a, b and λ are constants and $0 < \lambda < 1$. This equation and some more general forms are named after the pantographs on trains.

Solving DDEs

DDEs are mostly solved in a stepwise fashion with a principle called the method of steps. For instance, consider the DDE with a single delay

$$\frac{\mathrm{d}}{\mathrm{d}t}x(t) = f(x(t), x(t-\tau))$$

with given initial condition $\phi : [-\tau, 0] \rightarrow \mathbb{R}^n$. Then the solution on the interval $[0, \tau]$ is given by $\psi(t)$ which is the solution to the inhomogeneous initial value problem

$$\frac{\mathrm{d}}{\mathrm{d}t}\psi(t) = f(\psi(t), \phi(t-\tau))$$

with $\psi(0) = \phi(0)$. This can be continued for the successive intervals by using the solution to the previous interval as inhomogeneous term. In practice, the initial value problem is often solved numerically.

Reduction to ODE

In some cases, delay differential equations are equivalent to a system of ordinary differential equations.

- Example 1

  Consider an equation $\frac{\mathrm{d}}{\mathrm{d}t}x(t) = f\left(t, x(t), \int_{-\infty}^{0} x(t+\tau)e^{\lambda\tau}\,\mathrm{d}\tau\right)$.

  Introduce $y(t) = \int_{-\infty}^{0} x(t+\tau)e^{\lambda\tau}\,\mathrm{d}\tau$ to get a system of ODEs

  $$\frac{\mathrm{d}}{\mathrm{d}t}x(t) = f(t, x, y), \quad \frac{\mathrm{d}}{\mathrm{d}t}y(t) = x - \lambda y.$$

- Example 2

  An equation $\frac{\mathrm{d}}{\mathrm{d}t}x(t) = f\left(t, x(t), \int_{-\infty}^{0} x(t+\tau)\cos(\alpha\tau+\beta)\mathrm{d}\tau\right)$ is equivalent to $\frac{\mathrm{d}}{\mathrm{d}t}x(t) = f(t,x,y)$, $\frac{\mathrm{d}}{\mathrm{d}t}y(t) = \cos(\beta)x + \alpha z$, $\frac{\mathrm{d}}{\mathrm{d}t}z(t) = \sin(\beta)x - \alpha y$,

  where

$$y = \int_{-\infty}^{0} x(t+\tau)\cos(\alpha\tau+\beta)\mathrm{d}\tau, \quad z = \int_{-\infty}^{0} x(t+\tau)\sin(\alpha\tau+\beta)\mathrm{d}\tau.$$

The Characteristic Equation

Similar to ODEs, many properties of linear DDEs can be characterised and analysed using the *characteristic equation.* The characteristic equation associated with the linear DDE with discrete delays

$$\frac{\mathrm{d}}{\mathrm{d}t}x(t) = A_0 x(t) + A_1 x(t-\tau_1) + \cdots + A_m x(t-\tau_m)$$

is

$$\det(-\lambda I + A_0 + A_1 e^{-\tau_1\lambda} + \cdots + A_m e^{-\tau_m\lambda}) = 0$$

The roots λ of the characteristic equation are called characteristic roots or eigenvalues and the solution set is often referred to as the spectrum. Because of the exponential in the characteristic equation, the DDE has, unlike the ODE case, an infinite number of eigenvalues, making a spectral analysis more involved.

The spectrum does however have a some properties which can be exploited in the analysis. For instance, even though there are an infinite number of eigenvalues, there are only a finite number of eigenvalues to the right of any vertical line in the complex plane.

This characteristic equation is a nonlinear eigenproblem and there are many methods to compute the spectrum numerically.

In some special situations it is possible to solve the characteristic equation explicitly. Consider, for example, the following DDE:

$$\frac{\mathrm{d}}{\mathrm{d}t}x(t) = -x(t-1).$$

The characteristic equation is

$$-\lambda - e^{-\lambda} = 0.$$

There are an infinite number of solutions to this equation for complex λ. They are given by

$$\lambda = W_k(-1)$$

where W_k is the kth branch of the Lambert W function.

Solutions

In mathematics, in the study of differential equations, the Picard–Lindelöf theorem, Picard's existence theorem or Cauchy–Lipschitz theorem is an important theorem on existence and uniqueness of solutions to first-order equations with given initial conditions.

The theorem is named after Charles Émile Picard, Ernst Lindelöf, Rudolf Lipschitz and Augustin-Louis Cauchy.

Picard–Lindelöf Theorem

Consider the initial value problem

$$y'(t) = f(t, y(t)), \quad y(t_0) = y_0, \quad t \in [t_0 - \varepsilon, t_0 + \varepsilon].$$

Suppose f is Lipschitz continuous in y and continuous in t. Then, for some value $\epsilon > 0$, there exists a unique solution $y(t)$ to the initial value problem within the range $[t_0 - \epsilon, t_0 + \epsilon]$.

Proof Sketch

The proof relies on transforming the differential equation, and applying fixed-point theory.

By integrating both sides, any function satisfying the differential equation must also satisfy the integral equation

$$y(t) - y(t_0) = \int_{t_0}^{t} f(s, y(s))ds.$$

A simple proof of existence of the solution is obtained by successive approximations.

In this context, the method is known as Picard iteration.

Set

$$\varphi_0(t) = y_0$$

and

$$\varphi_{k+1}(t) = y_0 + \int_{t_0}^{t} f(s, \varphi_k(s))ds.$$

It can then be shown, by using the Banach fixed point theorem, that the sequence of "Picard iterates" φ_k is convergent and that the limit is a solution to the problem.

Exploiting the fact that the width of the interval where the local solution is defined is entirely determined by the Lipschitz constant of the function, one can assure global existence of the solution, i.e. the solution exists and is unique until it leaves the domain of definition of the ODE.

An application of Grönwall's lemma to $|\varphi(t) - \psi(t)|$, where φ and ψ are two solutions, shows that $\varphi(t) = \psi(t)$, thus proving the global uniqueness (the local uniqueness is a consequence of the uniqueness of the Banach fixed point).

Detailed Proof

Let $C_{a,b} = \overline{I_a(t_0)} \times \overline{B_b(y_0)}$ be the compact cylinder where f is defined, this is

$$t \in \overline{I_a(t_0)} = [t_0 - a, t_0 + a]$$

and

$$\overline{B_b(y_0)} = [y_0 - b, y_0 + b].$$

Let

$$M = \sup_{C_{a,b}} \| f \|,$$

this is, the maximum slope of the function in modulus. Finally, let L be the Lipschitz constant of with respect to the second variable.

We will proceed to apply Banach fixed point theorem using the metric on $C(I_a(t_0), B_b(y_0))$ induced by the uniform norm

$$\|\varphi_1\|_\infty = \sup_{t \in I_a} |\varphi(t)|.$$

We define an operator between two functional spaces of continuous functions, Picard's operator, as follows:

$$\Gamma : C(I_a(t_0), B_b(y_0)) \to C(I_a(t_0), B_b(y_0))$$

defined by:

$$\Gamma\varphi(t) = y_0 + \int_{t_0}^{t} f(s,\varphi(s))ds.$$

We impose that it is well-defined, in other words, that its image must be a function taking values on $B_b(y_0)$, or equivalently, that the norm of

$\Gamma\varphi(t) - y_0$ is less than b,

which can be restated as

$$\|\varphi_1\|_\infty \le b.$$

$$|\,|\Gamma\varphi(t) - y_0\,|\,| = |\,|\int_{t_0}^{t} f(s,\varphi(s))ds\,|\,| \le |\int_{t_0}^{t}|\,|$$

$$f(s,\varphi(s))\,|\,|ds| \le M\,|t - t_0| \le Ma \le b$$

The last step is the imposition, so we are impose the requirement

$$a < b / M.$$

Let us impose now the Picard's operator to be contractive under certain hypothesis over α that later on we will be able to omit.

Given two functions $\varphi_1, \varphi_2 \in C(I_\alpha(t_0), B_b(y_0))$, in order to apply the Banach fixed point theorem we want

$$\|\Gamma\varphi_1 - \Gamma\varphi_2\|_\infty \le q\|\varphi_1 - \varphi_2\|_\infty, \text{ for some } q < 1.$$

So let t be such that

$$\|\Gamma\varphi_1 - \Gamma\varphi_2\|_\infty = \|(\Gamma\varphi_1 - \Gamma\varphi_2)(t)\|$$

then using the definition of Γ

$$\|\ \|(\Gamma\varphi_1 - \Gamma\varphi_2)(t)\| =$$

$$\left\|\int_{t_0}^{t}(f(s,\varphi_1(s)) - f(s,\varphi_2(s))ds)\right\| \le \left|\int_{t_0}^{t}|\,|\,f(s,\varphi_1(s)) - f(s,\varphi_2(s))\,|\,|ds\right|.$$

$$\varphi_1 - \Gamma\varphi_2\|_\infty = \|(\Gamma\varphi_1 - \Gamma\varphi_2)(t)\|$$

Then since *f* is Lipschitz with respect to the second variable, we have that:

$$L\left|\int_{t_0}^{t}||\varphi_1(s)-\varphi_2(s)||ds\right|\leq La||\varphi_1-\varphi_2||.$$

This is contractive if $a < 1/L$. We have established that the Picard's operator is a contraction on the Banach spaces with the metric induced by the uniform norm. This allows us to apply the Banach fixed point theorem to conclude that the operator has a unique fixed point. In particular, there is a unique function

$$\varphi \in \mathcal{C}(I_a(t_0), B_b(y_0)) \text{ such that } \Gamma\varphi = \varphi.$$

This function is the unique solution of the initial value problem, valid on the interval I_a where a satisfies the condition

$$a < \min\{b/M, 1/L\}.$$

Other Existence Theorems

The Picard–Lindelöf theorem shows that the solution exists and that it is unique. The Peano existence theorem shows only existence, not uniqueness, but it assumes only that f is continuous in y, instead of Lipschitz continuous. For example, the right-hand side of the equation y
2 = $y^{1/3}$ with initial condition $y(0) = 0$ is continuous but not Lipschitz continuous. Indeed, the solution of this equation is not unique; two different solutions are given besides the trivial one $y(t) = 0$

$$y(t) = \pm\left(\tfrac{2}{3}t\right)^{3/2}.$$

Even more general is Carathéodory's existence theorem, which proves existence (in a more general sense) under weaker conditions on f.

Wronskian

In mathematics, the Wronskian is a determinant introduced by Jσzef Hoene-Wronski (1812) and named by Thomas Muir (1882Chapter XVIII).

It is used in the study of, differential equations, where it can sometimes be used to show that a set of solutions is linearly independent.

Definition

The Wronskian of two functions f and g is $W(f,g) = fg2 - gf\ 2$. More generally, for n real- or complex-valued functions $f_1,\ldots, f_n$, which are $n - 1$ times differentiable on an interval I, the Wronskian $W(f_1, \ldots, f_n)$ as a function on I is defined by

$$W(f_1,\ldots,f_n)(x) = \begin{vmatrix} f_1(x) & f_2(x) & \cdots & f_n(x) \\ f_{1'}(x) & f_{2'}(x) & \cdots & f_{n'}(x) \\ \vdots & \vdots & \ddots & \vdots \\ f_1^{(n-1)}(x) & f_2^{(n-1)}(x) & \cdots & f_n^{(n-1)}(x) \end{vmatrix}, \qquad x \in I.$$

That is, it is the determinant of the matrix constructed by placing the functions in the first row, the first derivative of each function in the second row, and so on through the $(n - 1)$st derivative, thus forming a square matrix sometimes called a fundamental matrix.

When the functions f_i are solutions of a linear differential equation, the Wronskian can be found explicitly using Abel's identity, even if the functions f_i are not known explicitly.

The Wronskian and Linear Independence

If the functions f_i are linearly dependent, then so are the columns of the Wronskian as differentiation is a linear operation, so the Wronskian vanishes. So the Wronskian can be used to show that a set of differentiable functions is linearly independent on an interval by showing that it does not vanish identically.

A common misconception is that $W = 0$ everywhere implies linear dependence, but Peano (1889) pointed out that the functions x^2 and $|x|x$ have continuous derivatives and their Wronskian vanishes everywhere, yet they are not linearly dependent in any neighbourhood of 0. There are several extra conditions which ensure that the vanishing of the Wronskian in an interval implies linear dependence.

Peano (1889) observed that if the functions are analytic, then the vanishing of the Wronskian in an interval implies that they are linearly dependent. Bochner (1901) gave several other conditions for the vanishing of the Wronskian to imply linear dependence; for example, if the Wronskian of n functions is identically zero and the n Wronskians of n–1 of them do not all vanish at any point then the functions are linearly dependent.

Wolsson (1989a) gave a more general condition that together with the vanishing of the Wronskian implies linear dependence.

Generalised Wronskians

For n functions of several variables, a generalised Wronskian is the determinant of an n by n matrix with entries $D_i(f_j)$ (with $0d-i<n$), where each D_i is some constant coefficient linear partial differential operator of order i. If the functions are linearly dependent then all generalised Wronskians vanish. As in the 1 variable case the converse is not true in general: if all generalised Wronskians vanish this does not imply that the functions are linearly dependent. However the converse is true in many special cases. For example, if the functions are polynomials and all generalised Wronskians vanish, then the functions are linearly dependent. Roth used this result about generalised Wronskians in his proof of Roth's theorem.

Phase Portrait

A phase portrait is a geometric representation of the trajectories of a dynamical system in the phase plane. Each set of initial conditions is represented by a different curve, or point.

Phase portraits are an invaluable tool in studying dynamical systems. They consist of a plot of typical trajectories in the state space. This reveals information such as whether an attractor, a repellor or limit cycle is present for the chosen parameter value. The concept of topological equivalence is important in classifying the behaviour of systems by specifying when two different phase portraits represent the same qualitative dynamic behaviour.

A phase portrait graph of a dynamical system depicts the system's trajectories (with arrows) and stable steady states (with dots) and unstable steady states (with circles) in a state space. The axes are of state variables.

Phase Space

In mathematics and physics, a phase space, introduced by Willard Gibbs in 1901, is a space in which all possible states of a system are represented, with each possible state of the system corresponding to one unique point in the phase space. For mechanical systems, the phase space usually consists of all possible values of position and momentum variables i.e. the cotangent space of configuration space.

A plot of position and momentum variables as a function of time is sometimes called a phase plot or a phase diagram. Phase diagram, however, is more usually reserved in the physical sciences for a diagram showing the various regions of stability of the thermodynamic phases of a chemical system, which consists of pressure, temperature, and composition.

In a phase space, every degree of freedom or parameter of the system is represented as an axis of a multidimensional space; a one-dimensional system is called a phase line, while a two-dimensional system is called a phase plane. For every possible state of the system, or allowed combination of values of the system's parameters, a point is plotted in the multidimensional space. Often this succession of plotted points is analogous to the system's state evolving over time. In the end, the phase diagram represents all that the system can be, and its shape can easily elucidate qualities of the system that might not be obvious otherwise. A phase space may contain very many dimensions. For instance, a gas containing many molecules may require a separate dimension for each particle's x, y and z positions and momenta as well as any number of other properties.

In classical mechanics, any choice of generalised coordinates q^i for the position (i.e. coordinates on configuration space) defines conjugate generalised momenta p_i which together define co-ordinates on phase space. More abstractly, in classical mechanics phase space is the cotangent space of configuration space, and in this interpretaton the procedure above expresses that a choice of local coordinates on configuration space induces a choice of natural local Darboux coordinates for the standard symplectic structure on a cotangent space.

The motion of an ensemble of systems in this space is studied by classical statistical mechanics. The local density of points in such systems obeys Liouville's Theorem, and so can be taken as constant. Within the context of a model system in classical mechanics, the phase space coordinates of the system at any given time are composed of all of the system's dynamical variables. Because of this, it is possible to calculate the state of the system at any given time in the future or the past, through integration of Hamilton's or Lagrange's equations of motion.

Examples

Low Dimensions: For simple systems, there may be as few as one or two degrees of freedom. One degree of freedom occurs when one has an autonomous ordinary differential equation in a single variable, $dy / dt = f(y)$, with the resulting one-dimensional system being called a phase line, and the qualitative behaviour of the system being immediately visible from the phase line. The simplest non-trivial examples are the exponential growth model/decay (one unstable/stable equilibrium) and the logistic growth model (two equilibria, one stable, one unstable).

The phase space of a two-dimensional system is called a phase plane, which occurs in classical mechanics for a single particle moving in one dimension, and where the two variables are position and velocity. In this case, a sketch of the phase portrait may give qualitative information about the dynamics of the system, such as the limit-cycle of the Van der Pol oscillator shown in the diagram. Here, the horizontal axis gives the position and vertical axis the velocity. As the system evolves, its state follows one of the lines (trajectories) on the phase diagram.

Chaos Theory

Classic examples of phase diagrams from chaos theory are:

- the Lorenz attractor
- population growth (i.e. Logistic map)
- parameter plane of complex quadratic polynomials with Mandelbrot set.

Quantum Mechanics

In quantum mechanics, the coordinates p and q of phase space normally become hermitian operators in a Hilbert space.

But they may alternatively retain their classical interpretation, provided functions of them compose in novel algebraic ways (through Groenewold's 1946 star product), consistent with the uncertainty principle of quantum mechanics. Every quantum mechanical observable corresponds to a unique function or distribution on phase space, and vice versa, as specified by Hermann Weyl (1927) and supplemented by John von Neumann (1931); Eugene Wigner (1932); and, in a grand synthesis, by H J Groenewold (1946). With J E Moyal (1949), these completed the foundations of the phase space formulation of quantum mechanics, a complete and logically autonomous reformulation of quantum mechanics. (Its modern abstractions include deformation quantisation and geometric quantisation.)

Expectation values in phase-space quantisation are obtained isomorphically to tracing operator observables with the density matrix in Hilbert space: they are obtained by phase-space integrals of observables, with the Wigner quasi-probability distribution effectively serving as a measure.

Thus, by expressing quantum mechanics in phase space (the same ambit as for classical mechanics), the Weyl map facilitates recognition of quantum mechanics as a deformation (generalisation) of classical mechanics, with deformation parameter ‘/*S*, where *S* is the action of the relevant process. (Other familiar deformations in physics involve the deformation of classical Newtonian into relativistic mechanics, with deformation parameter v/c; or the deformation of Newtonian gravity into General Relativity, with deformation parameter Schwarzschild-radius/characteristic-dimension.)

Classical expressions, observables, and operations (such as Poisson brackets) are modified by ‘-dependent quantum corrections, as the conventional commutative multiplication applying in classical mechanics is generalised to the noncommutative star-multiplication characterising quantum mechanics and underlying its uncertainty principle.

Thermodynamics and Statistical Mechanics

In thermodynamics and statistical mechanics contexts, the term phase space has two meanings: It is used in the same sense as in classical mechanics. If a thermodynamical system consists of *N* particles, then a point in the *6N*-dimensional phase space describes the dynamical state of every particle in that system, as each particle is associated with three position variables and three momentum variables. In this sense, a point in phase space is said to be a microstate of the system. *N* is typically on the order of Avogadro's number, thus describing the system at a microscopic level is often impractical. This leads us to the use of phase space in a different sense. The phase space can refer to the space that is parametrised by the *macroscopic* states of the system, such as pressure, temperature, etc. For instance, one can view the pressure-volume diagram or entropy-temperature diagrams as describing part of this phase space. A point in this phase space is correspondingly called a macrostate. There may easily be more than one microstate with the same macrostate. For example, for a fixed temperature, the system could have many dynamic configurations at the microscopic level. When used in this sense, a phase is a region of phase space where the system in question is in, for example, the liquid phase, or solid phase, etc.

Since there are many more microstates than macrostates, the phase space in the first sense is usually a manifold of much larger dimensions than the second sense. Clearly, many more parameters are required to register every detail of the system down to the molecular or atomic scale than to simply specify, say, the temperature or the pressure of the system.

Phase Integral

In classical statistical mechanics (continuous energies) the concept of phase space provides a classical analog to the partition function (sum over states) known as the phase integral. Instead of summing the Boltzmann factor over discretely spaced energy states (defined by appropriate integer quantum numbers for each degree of freedom) one may integrate over continuous phase space. Such integration essentially consists of two parts: integration of the momentum component of all degrees of freedom (momentum space) and integration of the position component of all degrees of freedom (configuration space). Once the phase integral is known, it may be related to the classical partition function by multiplication of a normalisation constant representing the number of quantum energy states per unit phase space. As shown in detail in, this normalisation constant is simply the inverse of Planck's constant raised to a power equal to the number of degrees of freedom for the system.

Chapter 9

Fast Multipole Method

The fast multipole method (FMM) is a mathematical technique that was developed to speed up the calculation of long-ranged forces in the n-body problem. It does this by expanding the system Green's function using a multipole expansion, which allows one to group sources that lie close together and treat them as if they are a single source.

The FMM has also been applied in accelerating the iterative solver in the method of moments (MOM) as applied to computational electromagnetics problems. The FMM was first introduced in this manner by Greengard and Rokhlin and is based on the multipole expansion of the vector Helmholtz equation. By treating the interactions between far-away basis functions using the FMM, the corresponding matrix elements do not need to be explicitly stored, resulting in a significant reduction in required memory.

If the FMM is then applied in a hierarchical manner, it can improve the complexity of matrix-vector products in an iterative solver from $O(N^2)$ to $O(N)$. This has expanded the area of applicability of the MOM to far greater problems than were previously possible.

The FMM, introduced by Rokhlin and Greengard, has been acclaimed as one of the top-ten algorithms of the 20th century. The FMM algorithm dramatically reduces the complexity of matrix-vector multiplication involving a certain type of dense matrix which can arise out of many physical systems. The FMM has also been applied for efficiently treating the Coulomb interaction in Hartree–Fock and density functional theory calculations in quantum chemistry.

Finite Difference

A finite difference is a mathematical expression of the form $f(x + b) - f(x + a)$. If a finite difference is divided by $b - a$, one gets a difference quotient. The approximation of derivatives by finite differences plays a central role in finite difference methods for the numerical solution of differential equations, especially boundary value problems. Recurrence relations can be written as difference equations by replacing iteration notation with finite differences.

Forward, Backward and Central Differences: Only three forms are commonly considered: forward, backward, and central differences.

A forward difference is an expression of the form

$$\Delta_h[f](x) = f(x+h) - f(x).$$

Depending on the application, the spacing h may be variable or constant. A backward difference uses the function values at x and $x - h$, instead of the values at $x + h$ and x:

$$\nabla_h[f](x) = f(x) - f(x-h).$$

Finally, the central difference is given by

$$\delta_h[f](x) = f(x+\tfrac{1}{2}h) - f(x-\tfrac{1}{2}h).$$

Relation with Derivatives: The derivative of a function f at a point x is defined by the limit

$$f'(x) = \lim_{h\to 0}\frac{f(x+h)-f(x)}{h}.$$

If h has a fixed (non-zero) value instead of approaching zero, then the right-hand side of the above equation would be written

$$\frac{f(x+h)-f(x)}{h} = \frac{\Delta_h[f](x)}{h}.$$

Hence, the forward difference divided by h approximates the derivative when h is small. The error in this approximation can be derived from Taylor's theorem. Assuming that f is continuously differentiable, the error is

$$\frac{\Delta_h[f](x)}{h} - f'(x) = O(h) \quad (h \to 0).$$

The same formula holds for the backward difference:

$$\frac{\nabla_h[f](x)}{h} - f'(x) = O(h).$$

However, the central difference yields a more accurate approximation. Its error is proportional to square of the spacing (if f is twice continuously differentiable):

$$\frac{\delta_h[f](x)}{h} - f'(x) = O(h^2).$$

The main problem with the central difference method, however, is that oscillating functions can yield zero derivative. If $f(nh)$ = 1 for n uneven, and $f(nh)$ = 2 for n even, then $f'(nh)$ = 0 if it is calculated with the central difference scheme. This is particularly troublesome if the domain of f is discrete.

Higher-order Differences

In an analogous way one can obtain finite difference approximations to higher order derivatives and differential operators. For example, by using the above central difference formula for $f'(x+h/2)$ and $f'(x-h/2)$ and applying a central difference formula for the derivative of f' at x, we obtain the central difference approximation of the second derivative of f:

2nd Order Central

$$f''(x) \approx \frac{\delta_h^2[f](x)}{h^2} = \frac{f(x+h) - 2f(x) + f(x-h)}{h^2}.$$

Similarly we can apply other differencing formulas in a recursive manner.

2nd Order Forward

$$f''(x) \approx \frac{\Delta_h^2[f](x)}{h^2} = \frac{f(x+2h) - 2f(x+h) + f(x)}{h^2}.$$

More generally, the nth-order forward, backward, and central differences are respectively given by:

$$\Delta_h^n[f](x) = \sum_{i=0}^{n} (-1)^i \binom{n}{i} f(x+(n-i)h),$$

$$\nabla_h^n[f](x) = \sum_{i=0}^{n} (-1)^i \binom{n}{i} f(x-ih),$$

$$\delta_h^n[f](x) = \sum_{i=0}^{n} (-1)^i \binom{n}{i} f\left(x + \left(\frac{n}{2} - i\right)h\right).$$

Note that the central difference will, for odd n, have h multiplied by non-integers. This is often a problem because it amounts to changing the interval of discretisation. The problem may be remedied taking the average of $\delta^n[f](x-h/2)$ and $\delta^n[f](x+h/2)$.

The relationship of these higher-order differences with the respective derivatives is very straightforward:

$$\frac{d^n f}{dx^n}(x) = \frac{\Delta_h^n[f](x)}{h^n} + O(h) = \frac{\nabla_h^n[f](x)}{h^n} + O(h) = \frac{\delta_h^n[f](x)}{h^n} + O(h^2).$$

Higher-order differences can also be used to construct better approximations. As mentioned above, the first-order difference approximates the first-order derivative up to a term of order h. However, the combination

$$\frac{\Delta_h[f](x) - \frac{1}{2}\Delta_h^2[f](x)}{h} = -\frac{f(x+2h) - 4f(x+h) + 3f(x)}{2h}$$

approximates $f'(x)$ up to a term of order h^2. This can be proven by expanding the above expression in Taylor series, or by using the calculus of finite differences, explained below.

If necessary, the finite difference can be centred about any point by mixing forward, backward, and central differences.

Arbitrarily Sized Kernels: Using a little linear algebra, one can fairly easily construct approximations, which sample an arbitrary number of points to the left and a (possibly different) number of points to the right of the centre point, for any order of derivative. This involves solving a linear system such that the Taylor expansion of the sum of those points, around the centre point, well approximates the Taylor expansion of the desired derivative.

This is useful for differentiating a function on a grid, where, as one approaches the edge of the grid, one must sample fewer and fewer points on one side.

The details are outlined in these notes.

Finite Difference Methods

An important application of finite differences is in numerical analysis, especially in numerical differential equations, which aim at the numerical solution of ordinary and partial differential equations respectively.

The idea is to replace the derivatives appearing in the differential equation by finite differences that approximate them. The resulting methods are called finite difference methods.

Common applications of the finite difference method are in computational science and engineering disciplines, such as thermal engineering, fluid mechanics, etc.

n-th difference

The nth forward difference of a function $f(x)$ is given by

$$\Delta^n[f](x) = \sum_{k=0}^{n} \binom{n}{k} (-1)^{n-k} f(x+k)$$

where $\binom{n}{k}$ is the binomial coefficient. Forward differences applied to a sequence are sometimes called the binomial transform of the sequence, and have a number of interesting combinatorial properties.

Forward differences may be evaluated using the Nörlund–Rice integral. The integral representation for these types of series is interesting because the integral can often be evaluated using asymptotic expansion or saddle-point techniques; by contrast, the forward difference series can be extremely hard to evaluate numerically, because the binomial coefficients grow rapidly for large n.

Finite Difference Methods for Option Pricing

Finite difference methods for option pricing are numerical methods used in mathematical finance for the valuation of options. Finite difference methods were first applied to option pricing by Eduardo Schwartz in 1977.

Finite difference methods can solve derivative pricing problems that have, in general, the same level of complexity as those problems solved by tree approaches, and are therefore usually employed only when other approaches are inappropriate. At the same time, like tree-based methods, this approach is limited in terms of the number of underlying variables, and for problems with multiple dimensions, Monte Carlo methods for option pricing are usually preferred.

The approach arises since the evolution of the option value can be modelled via a partial differential equation (PDE), as a function of (at least) time and price of underlying. Once in this form, a finite difference model can be derived, and the valuation obtained.

Here, essentially, the PDE is expressed in a discretised form, using finite differences, and the evolution in the option price is then modelled using a lattice with corresponding dimensions; time runs from 0 to maturity and price runs from 0 to a "high" value, such that the option is deeply in or out of the money.

The option is valued as follows:

- Maturity values are simply the difference between the exercise price of the option and the value of the underlying at each point.
- Values at the boundary prices are set based on moneyness or arbitrage bounds on option prices.
- Values at other lattice points are calculated recursively, starting at the time step preceding maturity and ending at time = 0. Here, using a technique such as Crank–Nicolson or the explicit method:
  1. the PDE is discretised per the technique chosen, such that the value at each lattice point is specified as a function of the value at later and adjacent points;
  2. the value at each point is then found using the technique in question.
- The value of the option today, where the underlying is at its spot price, (or at any time/price combination,) is then found by interpolation.

As above, these methods and tree-based methods are able to handle problems which are equivalent in complexity. In fact, when standard assumptions are applied it can be shown that the explicit technique encompasses the binomial and trinomial tree methods. Tree based methods, then, suitably parameterised, are a special case of the explicit finite difference method.

Finite Element Method in Structural Mechanics

The Finite element method (FEM) is a powerful technique originally developed for numerical solution of complex problems in structural mechanics, and it remains the method of choice for complex systems. In the FEM, the structural system is modelled by a set of appropriate finite elements interconnected at points called nodes. Elements may have physical properties such as thickness, coefficient of thermal expansion, density, Young's modulus, shear modulus and Poisson's ratio.

History

The origin of finite method can be traced to the matrix analysis of structures where the concept of displacement or stiffness matrix approach was introduced. Finite element concepts were developed based on engineering methods in 50s. The original works such as those by Argyris and Clough became foundation for today's finite element structural analysis methods. Earlier books such as by Zienkiewicz and more recent books such as by Yang give comprehensive summary of developments in finite-element structural analysis.

Element Properties

- Straight or curved one-dimensional elements with physical properties such as axial, bending, and torsional stiffnesses. This type of element is suitable for modelling cables, braces, trusses, beams, stiffeners, grids and frames. Straight elements usually have two nodes, one at each end, while curved elements will need at least three nodes including the end-nodes. The elements are positioned at the centroidal axis of the actual members.
- Two-dimensional elements for membrane action (plane stress, plane strain) and/or bending action (plates and shells). They may have a variety of shapes such as flat or curved triangles and quadrilaterals. Nodes are usually placed at the element corners and, if needed for higher accuracy, additional nodes can be placed along the element edges or even inside the element. The elements

are positioned at the mid-surface of the actual layer thickness.

- Torus-shaped elements for axisymmetric problems such as thin, thick plates, shells, and solids. The cross-section of the elements are similar to the previously described types: one-dimensional for thin plates and shells, and two-dimensional for solids, and thick plates and shells.
- Three-dimensional elements for modelling 3-D solids such as machine components, dams, embankments or soil masses. Common element shapes include tetrahedrals and hexahedrals. Nodes are placed at the vertexes and possibly in the element faces or within the element.

Element Interconnection and Displacement

The elements are interconnected only at the exterior nodes, and altogether they should cover the entire domain as accurately as possible. Nodes will have nodal (vector) displacements or degrees of freedom which may include translations, rotations, and for special applications, higher order derivatives of displacements. When the nodes displace, they will *drag* the elements along in a certain manner dictated by the element formulation. In other words, displacements of any points in the element will be interpolated from the nodal displacements, and this is the main reason for the approximate nature of the solution.

Practical Considerations

From the application point of view, it is important to model the system such that:

- Symmetry or anti-symmetry conditions are exploited in order to reduce the size of the domain.
- Displacement compatibility, including any required discontinuity, is ensured at the nodes, and preferably, along the element edges as well, particularly when adjacent elements are of different types, material or thickness. Compatibility of displacements of many nodes can usually be imposed via constraint relations—When such a feature is not available in the software package, a physical model that imposes the constraints may be used instead.

- Elements' behaviours capture the dominant actions of the actual system, both locally and globally.
- The element mesh is sufficiently fine in order to have acceptable accuracy. To assess accuracy, the mesh is refined until the important results shows little change. For higher accuracy, the aspect ratio of the elements should be as close to unity as possible, and smaller elements are used over the parts of higher stress gradient.
- Proper support constraints are imposed with special attention paid to nodes on symmetry axes.

Large scale commercial software packages often provide facilities for generating the mesh, graphical display of input and output, which greatly facilitate the verification of both input data and interpretation of the results.

Theoretical Overview of FEM-Displacement Formulation: From Elements to System to Solution

While the theory of FEM can be presented in different perspectives or emphases, its development for structural analysis follows the more traditional approach via the virtual work principle or the minimum total potential energy principle. The virtual work principle approach is more general as it is applicable to both linear and non-linear material behaviours. The principle of virtual displacements for the structural system expresses the mathematical identity of external and internal virtual work:

$$\text{External virtual work} = \int_V \delta\epsilon^T \sigma dV \tag{1}$$

The virtual internal work in the right-hand-side of the above equation may be found by summing the virtual work in the individual elements—This is the crucial step where we will need displacement functions written only for the small domain rather than over the entire system. As shown in the subsequent sections, Eq.(1) leads to the following governing equilibrium equation for the system:

$$R = Kr + R^o \tag{2}$$

where

R= vector of nodal forces, representing external forces applied to the system's nodes.

r= vector of system's nodal displacements, which will, by interpolation, yield displacements at any point of the finite element mesh.

R°= vector of equivalent nodal forces, representing all external effects other than the nodal forces which are already included in the preceding nodal force vector R. These external effects may include distributed or concentrated surface forces, body forces, thermal effects, initial stresses and strains.

K= system stiffness matrix, which will be established by *assembling* the *elements' stiffness matrices* :k^e.

Once the supports' constraints are accounted for, the nodal displacements are found by solving the system of linear equations (2), symbolically:

$$r = K^{-1}(R - R^o) \tag{3}$$

Subsequently, the strains and stresses in individual elements may be found as follows:

$$\epsilon = Bq \tag{4}$$

$$\sigma = E(\epsilon - \epsilon^o) + \sigma^o = E(Bq - \epsilon^o) + \sigma^o \tag{5}$$

where

q = vector of element's nodal displacements—a subset of the system displacement vector r that pertains to the element under consideration.

B = strain-displacement matrix that transforms nodal displacements q to strains at any point in the element.

E = elasticity matrix that transforms effective strains to stresses at any point in the element.

ϵ^o = vector of initial strains in the element.

σ^o = vector of initial stresses in the element.

By applying the virtual work equation (1) to the system, we can establish the element matrices B, k^e as well as the technique of assembling the system matrices R^o and K. Other matrices such as ϵ^o, σ^o, RandE can be directly set up from data input.

Interpolation or Shape Functions

Let q be the vector of nodal displacements of a typical element. The displacements at any point of the element may be found by interpolation functions as, symbolically:

$$u = Nq \tag{6}$$

where

u = vector of displacements at any point $\{x, y, z\}$ of the element.

N = matrix of *shape functions* serving as interpolation functions.

Equation (6) gives rise to other quantities of great interest:

- Virtual displacements consistent with virtual nodal displacements:

$$\delta u = N\delta q \tag{6b}$$

- Strains in the elements:

$$\epsilon = Du = DNq \tag{7}$$

where D= matrix of differential operators that convert displacements to strains using linear elasticity theory. Eq.(7) shows that matrix B in (4) is

$$B = DN \tag{8}$$

- Virtual strains consistent with element's virtual nodal displacements:

$$\delta\epsilon = B\delta q \tag{9}$$

Internal Virtual Work in a Typical Element

For a typical element of volume V^e, the internal virtual work due to virtual displacements is obtained by substitution of (5) and (9) into (1):

$$\text{Internal virtual work} = \int_{V^e} \delta\epsilon^T \sigma dV^e = \delta q^T \int_{V^e} B^T\left\{E(Bq - \epsilon^o) + \sigma^o\right\}dV^e \tag{10}$$

Finite Pointset Method

In applied mathematics, the Finite Pointset Method (FPM) is a method for the solution of the equations governing viscous fluid flows, including the effects of heat and mass transfer. FPM models problems in continuum mechanics. The method solves not only fluid flows, but also problems with elastic or plastic deformations. More generally spoken, FPM considers all the viscous as well as elastic/plastic stress tensors, and any mixture of it.

Description

FPM is a meshfree method. The basis of the computations is a point cloud, which represents the continuum or in other words a continuum domain (fluid or solid) is replaced by a discrete number of points, which are referred to as finite points. Each finite point carries all fluid information, like density, velocity, pressure, temperature. Finite points can move with fluid velocity (Lagrangian approach) or the flow information runs through the finite points if they are located constant in space (Eulerian approach). Also a mixture approach (Arbitrary Lagrangian Eulerian ALE) is possible. This is useful in case of using the Eulerian approach in combination with free surfaces or moving parts. Therefore, finite points themselves can be considered as geometrical grids of the fluid domain.

The finite point density is prescribed by a smoothing length defined locally. FPM does not use a rigid neighbourhood list for a certain finite point as it is required in a mesh based method. All neighbours are allowed to move and the neighbourhood list is re-computed each time step. Thus the simple idea of FPM is to use a dynamical discretisation method, but the finite points itself are carrying all fluid information.

This method has various advantages over grid-based techniques; for example, it can handle fluid domains, which change naturally, whereas grid based techniques require additional computational effort. The finite points have to completely cover the whole flow domain, i.e. the point cloud has to fulfill certain quality criteria (finite points are not allowed to form "holes" which means finite points have to find sufficiently numerous neighbours; also, finite points are not allowed to cluster; etc.).

The finite point cloud is a geometrical basis, which allows for a numerical formulation making FPM a general finite difference idea applied to continuum mechanics. That especially means, if the point cloud would reduce to a regular cubic point grid, then FPM would reduce to a classical finite difference method. The idea of general finite differences also means that FPM is not based on a weak formulation like Galerkin's approach. Rather, FPM is a strong formulation which models differential equations by direct approximation of the occurring differential operators. The method used is a moving least squares idea which was especially developed for FPM.

History

In order to overcome the disadvantages of the classical methods many approaches have been developed to simulate such flows (Hansbo 92, Harlow et al. 1965, Hirt et al. 1981, Kelecy et al. 1997, Kothe at el. 1992, Maronnier et al. 1999, Tiwari et al. 2000). A classical grid free Lagrangian method is Smoothed Particle Hydrodynamics (SPH), which was originally introduced to solve problems in astrophysics (Lucy 1977, Gingold et al. 1977).

It has since been extended to simulate the compressible Euler equations in fluid dynamics and applied to a wide range of problems. The method has also been extended to simulate inviscid incompressible free surface flows (Monaghan 94). The implementation of the boundary conditions is the main problem of the SPH method.

Another approach for solving fluid dynamic equations in a grid free framework is the moving least squares or least squares method (Belytschko et al. 1996, Dilts 1996, Kuhnert 99, Kuhnert 2000, Tiwari et al. 2001 and 2000). With this approach boundary conditions can be implemented in a natural way just by placing the finite points on boundaries and prescribing boundary conditions on them (Kuhnert 99). The robustness of this method is shown by the simulation results in the field of airbag deployment in car industry. Here, the membrane (or boundary) of the airbag changes very rapidly in time and takes a quite complicated shape (Kuhnert et al. 2000).

In (Tiwari et al. 2000) we have performed simulations of incompressible flows as the limit of the compressible Navier–Stokes equations with some stiff equation of state. This approach was first used in (Monaghan 92) to simulate incompressible free surface flows by SPH. The incompressible limit is obtained by choosing a very large speed of sound in the equation of state such that the Mach number becomes small. However the large value of the speed of sound restricts the time step to be very small due to the CFL-condition.

The projection method of Chorin (Chorin 68) is a widely used approach to solve problems governed by the incompressible Navier–Stokes equation in a grid based structure. In (Tiwari et al. 2001), this method has been applied to a grid free framework with the help of the weighted least squares method. The scheme gives accurate results for the incompressible Navier–Stokes equations. The occurring Poisson equation for the pressure field is solved by a grid free method. In (Tiwari et al. 2001), it has been shown that the Poisson equation can be solved accurately by this approach for any boundary conditions. The Poisson solver can be adapted to the weighted least squares approximation procedure with the condition that the Poisson equation and the boundary condition must be satisfied on each finite point. This is a local iteration procedure.

Courant–Friedrichs–Lewy Condition

In mathematics, the Courant–Friedrichs–Lewy condition (CFL condition) is a necessary condition for convergence while solving certain partial differential equations (usually hyperbolic PDEs) numerically by the method of finite differences. It arises when explicit time-marching schemes are used for the numerical solution. As a consequence, the time step must be less than a certain time in many explicit time-marching computer simulations, otherwise the simulation will produce incorrect results. The condition is named after Richard Courant, Kurt Friedrichs, and Hans Lewy who described it in their 1928 paper.

Heuristic Description

The information behind the condition is that, for example, if a wave is moving across a discrete spatial grid and we want to compute its amplitude at discrete time steps of equal length, then this length must be less than the time for the wave to travel to adjacent grid points. As a corollary, when the grid point separation is reduced, the upper limit for the time step also decreases. In essence, the numerical domain of dependence of any point in space and time (which data values in the initial conditions affect the numerical computed value at that point) must include the analytical domain of dependence (where in the initial conditions has an effect on the exact value of the solution at that point) in order to assure that the scheme can access the information required to form the solution.

The CFL Condition

In order to make a reasonably formally precise statement of the condition, it is necessary to define the following quantities

- Spatial coordinate: it is one of the coordinates of the physical space in which the problem is posed.
- Spatial dimension of the problem: it is the number n of spatial dimensions i.e. the number of spatial coordinates of the physical space where the problem is posed. Typical values are $n = 1$, $n = 2$ and $n = 3$.
- Time: it is the coordinate, acting as a parameter, which describes the evolution of the system, distinct from the spatial coordinates.

The spatial coordinates and the time are supposed to be discrete valued independent variables, whose minimal steps are called respectively the interval length and the time step: the CFL condition relates the length of the time step to a function interval lengths of each spatial variable.

Operatively, the CFL condition is commonly prescribed for those terms of the finite-difference approximation of general partial differential equations which model the advection phenomenon.

The One-dimensional Case

For one-dimensional case, the CFL has the following form:

$$C = \frac{u\Delta t}{\Delta x} \leq C_{max}$$

where the dimensionless number is called the Courant number,

- u is the velocity (whose dimension is Length/Time)
- Δt is the time step (whose dimension is Time)
- Δx is the length interval (whose dimension is Length).

The value of C_{max} changes with the method used to solve the discretised equation. If an explicit (time-marching) solver is used then typically $C_{max} = 1$. Implicit (matrix) solvers are usually less sensitive to numerical instability and so larger values of may be tolerated.

The Two and General n-dimensional Case

In the two-dimensional case, the CFL condition becomes

$$C = \frac{u_x \Delta t}{\Delta x} + \frac{u_y \Delta t}{\Delta y} \leq C_{max}$$

with obvious meaning of the symbols involved. By analogy with the two-dimensional case, the general CFL condition for the n-dimensional case is the following one:

$$C = \Delta t \sum_{i=1}^{n} \frac{u_{x_i}}{\Delta x_i} \leq C_{max}$$

Note that the interval length it is not required to be the same for each spatial variable Δx_i, $i=1,\dots,n$. This "degree of freedom" can be used in order to somewhat optimise the value of the time step for a particular problem, by varying the values of the different interval in order to keep it not too small.

Implications of the CFL Condition

The CFL Condition is Only a Necessary One: As already remarked, the CFL condition is a necessary condition, but may not be sufficient for the convergence of the finite-difference approximation of a given numerical problem. Thus, in order to establish the convergence of the finite-difference approximation, it is necessary to use other methods, which in turn could imply further limitations on the length of the time step and/or the lengths of the spatial intervals.

The CFL Condition can be a Very Strong Requirement: The CFL condition can be a very limiting constraint on the time step Δt : for example, in the finite-difference approximation of certain fourth-order nonlinear partial differential equations, it can have the following form

$$\frac{\Delta t}{(\Delta x)^4} < Cu$$

meaning that a decrease in the length interval Δx requires a fourth order decrease in the time step for the condition to be fulfilled. Therefore, when solving particularly stiff problems, efforts are often made to avoid the CFL condition, for example by using implicit methods. However, in a recent work, a modern dynamical systems approaches to modelling, based upon centre manifold theory, is demonstrated to provide theoretical support for the construction of non-traditional discretisations that automatically overcome the CFL restriction.

Finite-difference Time-domain Method

Finite-difference time-domain (FDTD) is a numerical analysis technique used for modelling computational electrodynamics (finding approximate solutions to the associated system of differential equations). Since it is a time-domain method, FDTD solutions can cover a wide frequency range with a single simulation run, and treat nonlinear material properties in a natural way.

The FDTD method belongs in the general class of grid-based differential time-domain numerical modelling methods. The time-dependent Maxwell's equations (in partial differential form) are discretised using central-difference approximations to the space and time partial derivatives. The resulting finite-difference equations are solved in either software or hardware in a leapfrog manner: the electric field vector components in a volume of space are solved at a given instant in time; then the magnetic field vector components in the same spatial volume are solved at the next instant in time; and the process is repeated over and over again until the desired transient or steady-state electromagnetic field behaviour is fully evolved.

The basic FDTD space grid and time-stepping algorithm trace back to a seminal 1966 paper by Kane Yee in IEEE Transactions on Antennas and Propagation. The descriptor "Finite-difference time-domain" and its corresponding "FDTD" acronym were originated by Allen Taflove in a 1980 paper in IEEE Transactions on Electromagnetic Compatibility.

Since about 1990, FDTD techniques have emerged as primary means to computationally model many scientific and engineering problems dealing with electromagnetic wave interactions with material structures. Current FDTD modelling applications range from near-DC (ultralow-frequency geophysics involving the entire Earth-ionosphere waveguide) through microwaves (radar signature technology, antennas, wireless communications devices, digital interconnects, biomedical imaging/treatment) to visible light (photonic crystals, nanoplasmonics, solitons, and biophotonics).

In 2006, an estimated 2,000 FDTD-related publications appeared in the science and engineering literature. At present (2008), there are at least 27 commercial/proprietary FDTD software vendors; 8 free-software/open-source-software FDTD projects; and 2 freeware/closed-source FDTD projects, some not for commercial use.

Workings of the FDTD Method

When Maxwell's differential equations are examined, it can be seen that the change in the E-field in time (the time derivative) is dependent on the change in the H-field across space (the curl). This results in the basic FDTD time-stepping relation that, at any point in space, the updated value of the E-field in time is dependent on the stored value of the E-field and the numerical curl of the local distribution of the H-field in space.

The H-field is time-stepped in a similar manner. At any point in space, the updated value of the H-field in time is dependent on the stored value of the H-field and the numerical curl of the local distribution of the E-field in space. Iterating the E-field and H-field updates results in a marching-in-time process wherein sampled-data analogs of the continuous electromagnetic waves under consideration propagate in a numerical grid stored in the computer memory.

This description holds true for 1-D, 2-D, and 3-D FDTD techniques. When multiple dimensions are considered, calculating the numerical curl can become complicated. Kane Yee's seminal 1966 paper proposed spatially staggering the vector components of the E-field and H-field about rectangular unit cells of a Cartesian computational grid so that each E-field vector component is located midway between a pair of H-field vector components, and conversely. This scheme, now known as a Yee lattice, has proven to be very robust, and remains at the core of many current FDTD software constructs.

Furthermore, Yee proposed a leapfrog scheme for marching in time wherein the E-field and H-field updates are staggered so that E-field updates are conducted midway during each time-step between successive H-field updates, and conversely. On the plus side, this explicit time-stepping scheme avoids the need to solve simultaneous equations, and furthermore yields dissipation-free numerical wave propagation.

On the minus side, this scheme mandates an upper bound on the time-step to ensure numerical stability. As a result, certain classes of simulations can require many thousands of time-steps for completion.

Using the FDTD Method

To implement an FDTD solution of Maxwell's equations, a computational domain must first be established. The computational domain is simply the physical region over which the simulation will be performed. The E and H fields are determined at every point in space within that computational domain.

The material of each cell within the computational domain must be specified. Typically, the material is either free-space (air), metal, or dielectric. Any material can be used as long as the permeability, permittivity, and conductivity are specified.

The permittivity of dispersive materials in tabular form cannot be directly substituted into the FDTD scheme. Instead, it can be approximated using multiple Debye, Drude, Lorentz or critical point terms. This approximation can be obtained using open fitting programs and does not necessary have physical meaning.

Once the computational domain and the grid materials are established, a source is specified. The source can be current on a wire, applied electric field or impinging plane wave. In the last case FDTD can be used to simulate light scattering from arbitrary shaped objects, planar periodic structures at various incident angles, and photonic band structure of infinite periodic structures.

Since the E and H fields are determined directly, the output of the simulation is usually the E or H field at a point or a series of points within the computational domain. The simulation evolves the E and H fields forward in time.

Processing may be done on the E and H fields returned by the simulation. Data processing may also occur while the simulation is ongoing. While the FDTD technique computes electromagnetic fields within a compact spatial region, scattered and/or radiated far fields can be obtained via near-to-far-field transformations.

Strengths of FDTD Modelling

Every modelling technique has strengths and weaknesses, and the FDTD method is no different.

FDTD is a versatile modelling technique used to solve Maxwell's equations. It is intuitive, so users can easily understand how to use it and know what to expect from a given model.

FDTD is a time-domain technique, and when a broadband pulse (such as a Gaussian pulse) is used as the source, then the response of the system over a wide range of frequencies can be obtained with a single simulation. This is useful in applications where resonant frequencies are not exactly known, or anytime that a broadband result is desired.

Since FDTD calculates the E and H fields everywhere in the computational domain as they evolve in time, it lends itself to providing animated displays of the electromagnetic field movement through the model. This type of display is useful in understanding what is going on in the model, and to help ensure that the model is working correctly.

The FDTD technique allows the user to specify the material at all points within the computational domain. A wide variety of linear and nonlinear dielectric and magnetic materials can be naturally and easily modelled.

FDTD allows the effects of apertures to be determined directly. Shielding effects can be found, and the fields both inside and outside a structure can be found directly or indirectly.

FDTD uses the E and H fields directly. Since most EMI/EMC modelling applications are interested in the E and H fields, it is convenient that no conversions must be made after the simulation has run to get these values.

Weaknesses of FDTD Modelling

Since FDTD requires that the entire computational domain be gridded, and the grid spatial discretisation must be sufficiently fine to resolve both the smallest electromagnetic wavelength and the smallest geometrical feature in the model, very large computational domains can be developed, which results in very long solution times. Models with long, thin features, (like wires) are difficult to model in FDTD because of the excessively large computational domain required.

There is no way to determine unique values for permittivity and permeability at a material interface.

Space and time steps must satisfy the CFL condition.

FDTD finds the E/H fields directly everywhere in the computational domain. If the field values at some distance are desired, it is likely that this distance will force the computational domain to be excessively large. Far-field extensions are available for FDTD, but require some amount of postprocessing.

Since FDTD simulations calculate the E and H fields at all points within the computational domain, the computational domain must be finite to permit its residence in the computer memory. In many cases this is achieved by inserting artificial boundaries into the simulation space. Care must be taken to minimise errors introduced by such boundaries. There are a number of available highly effective absorbing boundary conditions (ABCs) to simulate an infinite unbounded computational domain. Most modern FDTD implementations instead use a special absorbing "material", called a perfectly matched layer (PML) to implement absorbing boundaries.

Because FDTD is solved by propagating the fields forward in the time domain, the electromagnetic time response of the medium must be modelled explicitly. For an arbitrary response, this involves a computationally expensive time convolution, although in most cases the time response of the medium (or Dispersion (optics)) can be adequately and simply modelled using either the recursive convolution (RC) technique, the auxiliary differential equation (ADE) technique, or the Z-transform technique. An alternative way of solving Maxwell's equations that can treat arbitrary dispersion easily is the Pseudospectral Spatial-Domain method (PSSD), which instead propagates the fields forward in space.

Grid Truncation Techniques for Open-region FDTD Modelling Problems

The most commonly used grid truncation techniques for open-region FDTD modelling problems are the Mur absorbing boundary condition (ABC), the Liao ABC, and various perfectly matched layer (PML) formulations. The Mur and Liao techniques are simpler than PML. However, PML (which is technically an absorbing region rather than a boundary condition *per se*) can provide orders-of-magnitude lower reflections. The PML concept was introduced by J.-P. Berenger in a seminal 1994 paper in the Journal of Computational Physics. Since 1994, Berenger's original split-field implementation has been modified and extended to the uniaxial PML (UPML), the convolutional PML (CPML), and the higher-order PML. The latter two PML formulations have increased ability to absorb evanescent waves, and therefore can in principle be placed closer to a simulated scattering or radiating structure than Berenger's original formulation.

To reduce undesired numerical reflection from the PML additional back absorbing layers technique can be used.

History of FDTD Techniques and Applications for Maxwell's Equations

An appreciation of the basis, technical development, and possible future of FDTD numerical techniques for Maxwell's equations can be developed by first considering their history. The following lists some of the key publications in this area, starting with Yee's seminal Paper #1 (1966), which has over 4,000 citations according to the ISI Web of Science.

Table: *Partial chronology of FDTD techniques and applications for Maxwell's equations.*

year	*event*
1966	Yee described the basis of the FDTD numerical technique for solving Maxwell's curl equations directly in the time domain on a space grid.

Contd...

year	*event*
1975	Taflove and Brodwin reported the correct numerical stability criterion for Yee's algorithm; the first sinusoidal steady-state FDTD solutions of two- and three-dimensional electromagnetic wave interactions with material structures; and the first bioelectromagnetics models.
1977	Holland and Kunz & Lee applied Yee's algorithm to EMP problems.
1980	Taflove coined the FDTD acronym and published the first validated FDTD models of sinusoidal steady-state electromagnetic wave penetration into a three-dimensional metal cavity.
1981	Mur published the first numerically stable, second-order accurate, absorbing boundary condition (ABC) for Yee's grid.
1982–83	Taflove and Umashankar developed the first FDTD electromagnetic wave scattering models computing sinusoidal steady-state near-fields, far-fields, and radar cross-section for two- and three-dimensional structures.
1984	Liao *et al* reported an improved ABC based upon space-time extrapolation of the field adjacent to the outer grid boundary.1985 Gwarek introduced the lumped equivalent circuit formulation of FDTD.
1986	Choi and Hoefer published the first FDTD simulation of waveguide structures.
1987–88	Kriegsmann *et al* and Moore *et al* published the first articles on ABC theory in *IEEE Transactions on Antennas and Propagation*.
1987–88, 1992	Contour-path subcell techniques were introduced by Umashankar *et al* to permit FDTD modelling of thin wires and wire bundles, by Taflove *et al* to model penetration through cracks in conducting screens, and by Jurgens *et al* to conformally model the surface of a smoothly curved scatterer.1988 Sullivan *et al* published the first 3-D FDTD model of sinusoidal steady-state electromagnetic wave absorption by a complete human body.

Contd...

year	***event***
1988	FDTD modelling of microstrips was introduced by Zhang *et al.*
1990–91	FDTD modelling of frequency-dependent dielectric permittivity was introduced by Kashiwa and Fukai, Luebbers *et al*, and Joseph *et al.*
1990–91	FDTD modelling of antennas was introduced by Maloney *et al*, Katz *et al*, and Tirkas and Balanis.
1990	FDTD modelling of picosecond optoelectronic switches was introduced by Sano and Shibata, and El-Ghazaly *et al.*
1992–94	FDTD modelling of the propagation of optical pulses in nonlinear dispersive media was introduced, including the first temporal solitons in one dimension by Goorjian and Taflove; beam self-focusing by Ziolkowski and Judkins; the first temporal solitons in two dimensions by Joseph *et al*; and the first spatial solitons in two dimensions by Joseph and Taflove.
1992	FDTD modelling of lumped electronic circuit elements was introduced by Sui *et al.*
1993	Toland *et al* published the first FDTD models of gain devices (tunnel diodes and Gunn diodes) exciting cavities and antennas.
1994	Thomas *et al* introduced a Norton's equivalent circuit for the FDTD space lattice, which permits the SPICE circuit analysis tool to implement accurate subgrid models of nonlinear electronic components or complete circuits embedded within the lattice.
1994	Berenger introduced the highly effective, perfectly matched layer (PML) ABC for two-dimensional FDTD grids, which was extended to three dimensions by Katz *et al*, and to dispersive waveguide terminations by Reuter *et al.*
1995–96	Sacks *et al* and Gedney introduced a physically realisable, uniaxial perfectly matched layer (UPML) ABC.
1997	Liu introduced the pseudospectral time-domain (PSTD) method, which permits extremely coarse spatial sampling of the electromagnetic field at the Nyquist limit.

Contd...

year	*event*
1997	Ramahi introduced the complementary operators method (COM) to implement highly effective analytical ABCs.
1998	Maloney and Kesler introduced several novel means to analyse periodic structures in the FDTD space lattice.
1998	Nagra and York introduced a hybrid FDTD-quantum mechanics model of electromagnetic wave interactions with materials having electrons transitioning between multiple energy levels.
1998	Hagness *et al* introduced FDTD modelling of the detection of breast cancer using ultrawideband radar techniques.
1999	Schneider and Wagner introduced a comprehensive analysis of FDTD grid dispersion based upon complex wavenumbers.
2000–01	Zheng, Chen, and Zhang introduced the first three-dimensional alternating-direction implicit (ADI) FDTD algorithm with provable unconditional numerical stability.
2000	Roden and Gedney introduced the advanced convolutional PML (CPML) ABC.
2000	Rylander and Bondeson introduced a provably stable FDTD - finite-element time-domain hybrid technique.
2002	Hayakawa *et al* and Simpson and Taflove independently introduced FDTD modelling of the global Earth-ionosphere waveguide for extremely low-frequency geophysical phenomena.
2003	DeRaedt introduced the unconditionally stable, "one-step" FDTD technique.
2008	Ahmed, Chua, Li and Chen introduced the three-dimensional locally one dimensional (LOD)FDTD method and proved unconditional numerical stability.Popularity

Interest in FDTD Maxwell's equations solvers has increased nearly exponentially over the past 20 years. Increasingly, engineers and scientists in nontraditional electromagnetics-related areas such as photonics and nanotechnology have become aware of the power of FDTD techniques. As shown in the figure on the right, an estimated 2,000 FDTD-related publications appeared in the science and engineering literature in 2006, as opposed to fewer than 10 as recently as 1985. The current rate of growth (based upon a study of ISI Web of Science data) is approximately 5:1 over the period 1995 to 2006. Furthermore, the descriptor "finite difference time domain" has become widely used, having appeared in this exact form in 43,900 articles as of August 7, 2011, according to Google Scholar.

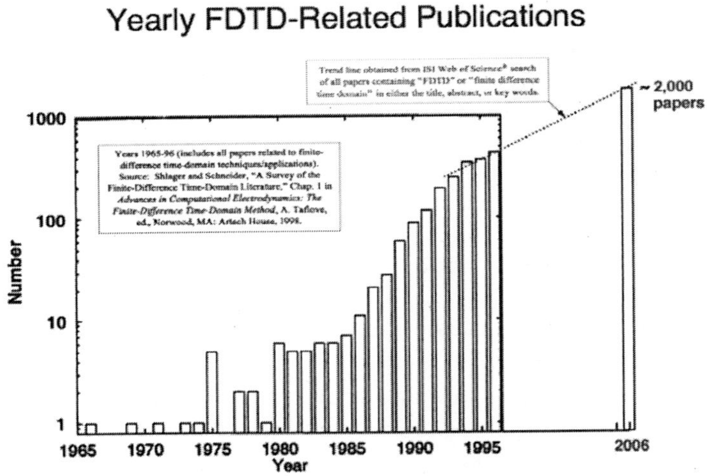

Notwithstanding both the general increase in academic publication throughput during the same period and the overall expansion of interest in all Computational electromagnetics (CEM) techniques, there are seven primary reasons for the tremendous expansion of interest in FDTD computational solution approaches for Maxwell's equations:

1. FDTD uses no linear algebra. Being a fully explicit computation, FDTD avoids the difficulties with linear algebra that limit the size of frequency-domain integral-equation and finite-element electromagnetics models to generally fewer than 10^9 electromagnetic field unknowns. FDTD models with as many as 10^9 field unknowns have been run; there is no intrinsic upper bound to this number.

2. FDTD is accurate and robust. The sources of error in FDTD calculations are well understood, and can be bounded to permit accurate models for a very large variety of electromagnetic wave interaction problems.
3. FDTD treats impulsive behaviour naturally. Being a time-domain technique, FDTD directly calculates the impulse response of an electromagnetic system. Therefore, a single FDTD simulation can provide either ultrawideband temporal waveforms or the sinusoidal steady-state response at any frequency within the excitation spectrum.
4. FDTD treats nonlinear behaviour naturally. Being a time-domain technique, FDTD directly calculates the nonlinear response of an electromagnetic system. This allows natural hybriding of FDTD with sets of auxiliary differential equations that describe nonlinearities from either the classical or semi-classical standpoint. One research frontier is the development of hybrid algorithms which join FDTD classical electrodynamics models with phenomena arising from quantum electrodynamics, especially vacuum fluctuations, such as the Casimir effect.
5. FDTD is a systematic approach. With FDTD, specifying a new structure to be modelled is reduced to a problem of mesh generation rather than the potentially complex reformulation of an integral equation. For example, FDTD requires no calculation of structure-dependent Green functions.
6. Parallel-processing computer architectures have come to dominate supercomputing. FDTD scales with high efficiency on parallel-processing CPU-based computers, and extremely well on recently developed GPU-based accelerator technology.
7. Computer visualisation capabilities are increasing rapidly. While this trend positively influences all numerical techniques, it is of particular advantage to FDTD methods, which generate time-marched arrays of field quantities suitable for use in colour videos to illustrate the field dynamics.

Taflove has argued that these factors combine to suggest that FDTD will remain one of the dominant computational electrodynamics techniques (as well as potentially other multiphysics problems).

High-resolution Scheme

High-resolution schemes are used in the numerical solution of partial differential equations where high accuracy is required in the presence of shocks or discontinuities. They have the following properties:

- Second or higher order spatial accuracy is obtained in smooth parts of the solution.
- Solutions are free from spurious oscillations or wiggles.
- High accuracy is obtained around shocks and discontinuities.
- The number of mesh points containing the wave is small compared with a first-order scheme with similar accuracy.

High-resolution schemes often use flux/slope limiters to limit the gradient around shocks or discontinuities. A particularly successful high-resolution scheme is the MUSCL scheme which uses state extrapolation and limiters to achieve good accuracy.

Lax Equivalence Theorem

In numerical analysis, the Lax equivalence theorem is the fundamental theorem in the analysis of finite difference methods for the numerical solution of partial differential equations. It states that for a consistent finite difference method for a well-posed linear initial value problem, the method is convergent if and only if it is stable.

The importance of the theorem is that while convergence of the solution of the finite difference method to the solution of the partial differential equation is what is desired, it is ordinarily difficult to establish because the numerical method is defined by a recurrence relation while the differential equation involves a differentiable function.

However, consistency—the requirement that the finite difference method approximate the correct partial differential equation—is straightforward to verify, and stability is typically much easier to show than convergence (and would be needed in any event to show that round-off error will not destroy the computation). Hence convergence is usually shown via the Lax equivalence theorem.

Stability in this context means that a matrix norm of the matrix used in the iteration is at most unity, called (practical) Lax-Richtmyer stability.

Often a von Neumann stability analysis is substituted for convenience, although von Neumann stability only implies Lax-Richtmyer stability in certain cases.

This theorem is due to Peter Lax. It is sometimes called the Lax–Richtmyer theorem, after Peter Lax and Robert D. Richtmyer.

Lax–Wendroff Method

The Lax–Wendroff method, named after Peter Lax and Burton Wendroff, is a numerical method for the solution of hyperbolic partial differential equations, based on finite differences. It is second-order accurate in both space and time.

Suppose one has an equation of the following form:

$$\frac{\partial f(x,t)}{\partial t} = \frac{\partial g(f(x,t))}{\partial x}$$

where x and t are independent variables, and the initial state, $f(x, 0)$ is given.

The first step in the Lax–Wendroff method calculates values for $f(x, t)$ at half time steps, $t_{n+1/2}$ and half grid points, $x_{i+1/2}$.

In the second step values at t_{n+1} are calculated using the data for t_n and $t_{n+1/2}$.

First (Lax) step:

$$\frac{f_{i+1/2}^{n+1/2} - \frac{f_i^n + f_{i+1}^n}{2}}{(1/2) * \Delta t} = \frac{g_{i+1}^n - g_i^n}{\Delta x}.$$

Second step:

$$\frac{f_i^{n+1} - f_i^n}{\Delta t} = \frac{g_{i+1/2}^{n+1/2} - g_{i-1/2}^{n+1/2}}{\Delta x}.$$

This method can be further applied to some systems of partial differential equations.

Lax–Wendroff Theorem

In computational mathematics, the Lax–Wendroff theorem, named after Peter Lax and Burton Wendroff, states that if a conservative numerical scheme for a hyperbolic system of conservation laws converges, then it converges towards a weak solution.

Leapfrog Integration

Leapfrog integration is a simple method for numerically integrating differential equations of the form

$$\ddot{x} = F(x)$$

or equivalently of the form

$$\dot{v} = F(x), \dot{x} \equiv v$$

particularly in the case of a dynamical system of classical mechanics. Such problems often take the form

$$\ddot{x} = -\nabla V(x),$$

with energy function

$$E(x,v) = \tfrac{1}{2} | v |^2 + V(x)$$

where V is the potential energy of the system. The method is known by different names in different disciplines. In particular, it is similar to the Velocity Verlet method, which is a variant of Verlet integration. Leapfrog integration is equivalent to updating positions $x(t)$ and velocities $v(t) = \dot{x}(t)$ at interleaved time points, staggered in such a way that they 'leapfrog' over each other. For example, the position is updated at integer time steps and the velocity is updated at integer-plus-a-half time steps.

Leapfrog integration is a second order method, in contrast to Euler integration, which is only first order, yet requires the same number of function evaluations per step. Unlike Euler integration, it is stable for oscillatory motion, as long as the time-step, $\Delta t < 1/\omega$

In leapfrog integration, the equations for updating position and velocity are

$$\begin{aligned} x_i &= x_{i-1} + v_{i-1/2}\Delta t, \\ [0.4em]a_i &= F(x_i) \\ [0.4em]v_{i+1/2} &= v_{i-1/2} + a_i \Delta t, \end{aligned}$$

where x_i is position at step i, $i+1/2$ is the velocity, or first derivative of x, at step $i+1/2$, $a_i = F(x_i)$ is the acceleration, or second derivative of, at step and is the size of each time step. These equations can be expressed in a form which gives velocity at integer steps as

$$\begin{aligned} x_{i+1} &= x_i + v_i \Delta t + \tfrac{1}{2} a_i \Delta t^2, \\ [0.4em]v_{i+1} &= v_i + \tfrac{1}{2}(a_i + a_{i+1})\Delta t. \end{aligned}$$

One use of this equation is in gravity simulations, since in that case the acceleration depends only on the positions of the gravitating masses, although higher order integrators (such as Runge–Kutta methods) are more frequently used.

There are two primary strengths to Leapfrog integration when applied to mechanics problems. The first is the time-reversibility of the Leapfrog method. One can integrate forward n steps, and then reverse the direction of integration and integrate backwards n steps to arrive at the same starting position.

The second strength of Leapfrog integration is its symplectic nature, which implies that it conserves the (slightly modified) energy of dynamical systems. This is especially useful when computing orbital dynamics, as other integration schemes, such as the Runge-Kutta method, do not conserve energy and allow the system to drift substantially over time.

Because of its time-reversibility, and because it is a symplectic integrator, leapfrog integration is also used in Hamiltonian Monte Carlo, a method for drawing random samples from a probability distribution whose overall normalisation is unknown.

MacCormack Method

In computational fluid dynamics, the MacCormack method is a widely used discretisation scheme for the numerical solution of hyperbolic partial differential equations. This second-order finite difference method is introduced by Robert W. MacCormack in 1969. The MacCormack method is very elegant and easy to understand and program.

The Algorithm

The MacCormack method is a variation of the two-step Lax–Wendroff scheme but is much simpler in application. To illustrate the algorithm, consider the following first order hyperbolic equation

$$\frac{\partial u}{\partial t}+a\frac{\partial u}{\partial x}=0.$$

The application of MacCormack method to the above equation proceeds in two steps; a *predictor step* which is followed by a *corrector step*.

Predictor step: In the predictor step, a "provisional" value of u at time level $n+1$ (denoted by $u_i^{\overline{n+1}}$) is estimated as follows

$$u_i^{\overline{n+1}}=u_i^n-a\frac{\Delta t}{\Delta x}\left(u_{i+1}^n-u_i^n\right)$$

It may be noted that the above equation is obtained by replacing the spatial and temporal derivatives in the previous first order hyperbolic equation using forward differences.

Corrector step: In the corrector step, the predicted value $u_i^{\overline{n+1}}$ is corrected according to the equation

$$u_i^{n+1}=u_i^{n+1/2}-a\frac{\Delta t}{2\Delta x}\left(u_i^{\overline{n+1}}-u_{i-1}^{\overline{n+1}}\right)$$

Note that the corrector step uses backward finite difference approximations for spatial derivative. Note also that the time-step used in the corrector step is $\Delta t/2$ in contrast to the Δt used in the predictor step.

Replacing the $u_i^{n+1/2}$ term by the temporal average

$$u_i^{n+1/2}=\frac{u_i^n+u_i^{\overline{n+1}}}{2}$$

to obtain the corrector step as

$$u_i^{n+1}=\frac{u_i^n+u_i^{\overline{n+1}}}{2}-a\frac{\Delta t}{2\Delta x}\left(u_i^{\overline{n+1}}-u_{i-1}^{\overline{n+1}}\right)$$

Some Remarks

The MacCormack method is well suited for nonlinear equations (Inviscid Burgers equation, Euler equations, etc.) The order of differencing can be reversed for the time step (i.e., forward/backward followed by backward/forward).

For nonlinear equations, this procedure provides the best results. For linear equations, the MacCormack scheme is equivalent to the Lax–Wendroff scheme.

Unlike first-order upwind scheme, the MacCormack does not introduce diffusive errors in the solution. However, it is known to introduce dispersive errors (Gibbs phenomenon) in the region where the gradient is high.

Chapter 10

Integral Transform

In mathematics, an integral transform is any transform T of the following form:

$$(Tf)(u) = \int_{t_1}^{t_2} K(t,u)\, f(t)\, dt$$

The input of this transform is a function f, and the output is another function Tf. An integral transform is a particular kind of mathematical operator.

There are numerous useful integral transforms. Each is specified by a choice of the function K of two variables, the kernel function or nucleus of the transform.

Some kernels have an associated *inverse kernel* $K^{-1}(u, t)$ which (roughly speaking) yields an inverse transform:

$$f(t) = \int_{u_1}^{u_2} K^{-1}(u,t)(Tf(u))\, du$$

A *symmetric kernel* is one that is unchanged when the two variables are permuted.

Motivation

Mathematical notation aside, the motivation behind integral transforms is easy to understand. There are many classes of problems that are difficult to solve—or at least quite unwieldy algebraically—in their original representations. An integral transform "maps" an equation from its original "domain" into another domain. Manipulating and solving the equation in the target domain can be much easier than manipulation and solution in the original domain. The solution is then mapped back to the original domain with the inverse of the integral transform.

History

The precursor of the transforms were the Fourier series to express functions in finite intervals. Later the Fourier transform was developed to remove the requirement of finite intervals.

Using the Fourier series, just about any practical function of time (the voltage across the terminals of an electronic device for example) can be represented as a sum of sines and cosines, each suitably scaled (multiplied by a constant factor), shifted (advanced or retarded in time) and "squeezed" or "stretched" (increasing or decreasing the frequency). The sines and cosines in the Fourier series are an example of an orthonormal basis.

Importance of Orthogonality

The individual basis functions have to be orthogonal. That is, the product of two dissimilar basis functions—integrated over their domain—must be zero. An integral transform, in actuality, just changes the representation of a function from one orthogonal basis to another.

Each point in the representation of the transformed function in the image of the transform corresponds to the contribution of a given orthogonal basis function to the expansion.

The process of expanding a function from its "standard" representation to a sum of a number of orthonormal basis functions, suitably scaled and shifted, is termed "spectral factorisation."

This is similar in concept to the description of a point in space in terms of three discrete components, namely, its x, y, and z coordinates. Each axis correlates only to itself and nothing to the other orthogonal axes.

Note the terminological consistency: the determination of the amount by which an individual orthonormal basis function must be scaled in the spectral factorisation of a function, F, is termed the "projection" of F onto that basis function.

The normal Cartesian graph *per se* of a function can be thought of as an orthonormal expansion. Indeed, each point just reflects the contribution of a given orthonormal basis function to the sum.

In this way, the graph of a continuous real-valued function in the plane corresponds to an *infinite* set of basis functions; if the number of basis functions were finite, the curve would consist of a discrete set of points rather than a continuous contour.

Usage Example

As an example of an application of integral transforms, consider the Laplace transform. This is a technique that maps differential or integro-differential equations in the "time" domain into polynomial equations in what is termed the "complex frequency" domain. (Complex frequency is similar to actual, physical frequency but rather more general.

Specifically, the imaginary component ω of the complex frequency $s = -\sigma + i\omega$ corresponds to the usual concept of frequency, *viz.*, the speed at which a sinusoid cycles, whereas the real component σ of the complex frequency corresponds to the degree of "damping".)

The equation cast in terms of complex frequency is readily solved in the complex frequency domain (roots of the polynomial equations in the complex frequency domain correspond to eigenvalues in the time domain), leading to a "solution" formulated in the frequency domain.

Employing the inverse transform, *i.e.*, the inverse procedure of the original Laplace transform, one obtains a time-domain solution. In this example, polynomials in the complex frequency domain (typically occurring in the denominator) correspond to power series in the time domain, while axial shifts in the complex frequency domain correspond to damping by decaying exponentials in the time domain.

The Laplace transform finds wide application in physics and particularly in electrical engineering, where the characteristic equations that describe the behaviour of an electric circuit in the complex frequency domain correspond to linear combinations of exponentially damped, scaled, and time-shifted sinusoids in the time domain. Other integral transforms find special applicability within other scientific and mathematical disciplines.

Different Domains

Here integral transforms are defined for functions on the real numbers, but they can be defined more generally for functions on a group.

- If instead one uses functions on the circle (periodic functions), integration kernels are then biperiodic functions; convolution by functions on the circle yields circular convolution.
- If one uses functions on the cyclic group of order n (C_n or $\mathbf{Z}/n\mathbf{Z}$), one obtains $n \times n$ matrices as integration kernels; convolution corresponds to circulant matrices.

General Theory

Although the properties of integral transforms vary widely, they have some properties in common. For example, every integral transform is a linear operator, since the integral is a linear operator, and in fact if the kernel is allowed to be a generalised function then all linear operators are integral transforms (a properly formulated version of this statement is the Schwartz kernel theorem).

The general theory of such integral equations is known as Fredholm theory. In this theory, the kernel is understood to be a compact operator acting on a Banach space of functions. Depending on the situation, the kernel is then variously referred to as the Fredholm operator, the nuclear operator or the Fredholm kernel.

Euler Method

In mathematics and computational science, the Euler method is a first-order numerical procedure for solving ordinary differential equations (ODEs) with a given initial value. It is the most basic explicit method for numerical integration of ordinary differential equations and is the simplest Runge–Kutta method.

The Euler method is named after Leonhard Euler, who treated it in his book *Institutionum calculi integralis* (published

1768–70). The Euler method is a first-order method, which means that the local error (error per step) is proportional to the square of the step size, and the global error (error at a given time) is proportional to the step size. It also suffers from stability problems. For these reasons, the Euler method is not often used in practice. It serves as the basis to construct more complicated methods.

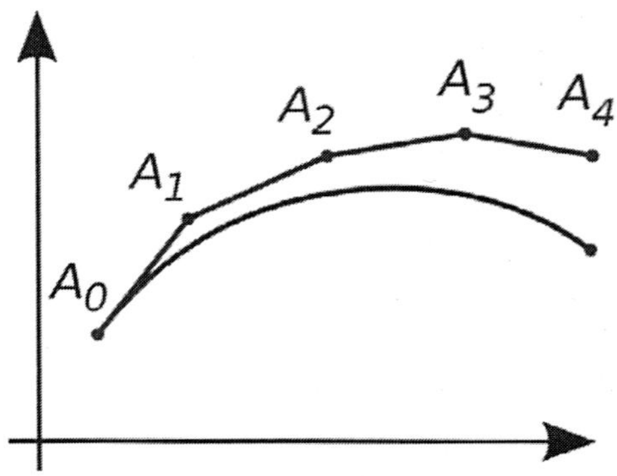

Figure: *Illustration of the Euler method. The unknown curve is in blue, and its polygonal approximation is in red.*

Formulation of the Method

Suppose that we want to approximate the solution of the initial value problem

$$y'(t) = f(t, y(t)), \qquad y(t_0) = y_0.$$

Choose a value h for the size of every step and set $t_n = t_0 + nh$ Now, one step of the Euler method from t_n to $t_{n+1} = t_n + h$ is

$$y_{n+1} = y_n + hf(t_n, y_n).$$

The value of y_n is an approximation of the solution to the ODE at time $t_n : y_n \approx y(t_n)$. The Euler method is explicit, i.e. the solution y_{n+1} is an explicit function of y_i for $i \le n$. While the Euler method integrates a first-order ODE, any ODE of order N can be represented as a first-order ODE: to treat the equation

$$y^{(N)}(t) = f(t, y(t), y'(t), \ldots, y^{(N-1)}(t)),$$

we introduce auxiliary variables

$$z_1(t) = y(t), z_2(t) = y'(t), \ldots, z_N(t) = y^{(N-1)}(t)$$

and obtain the equivalent equation

$$\mathbf{z}'(t) = \begin{pmatrix} z_{1'}(t) \\ \vdots \\ z_{N-1'}(t) \\ z_{N'}(t) \end{pmatrix} = \begin{pmatrix} y'(t) \\ \vdots \\ y^{(N-1)}(t) \\ y^{(N)}(t) \end{pmatrix} = \begin{pmatrix} z_2(t) \\ \vdots \\ z_N(t) \\ f(t, z_1(t), \ldots, z_N(t)) \end{pmatrix}$$

This is a first-order system in the variable $\mathbf{z}(t)$ and can be handled by Euler's method or, in fact, by any other scheme for first-order systems.

Example

Given the initial value problem

$$y' = y, \quad y(0) = 1,$$

we would like to use the Euler method to approximate $y(4)$.

Derivation

The Euler method can be derived in a number of ways. Firstly, there is the geometrical description mentioned above.

Another possibility is to consider the Taylor expansion of the function y around t_0:

$$y(t_0 + h) = y(t_0) + hy'(t_0) + \frac{1}{2}h^2 y''(t_0) + O(h^3).$$

The differential equation states that $y' = f(t, y)$. If this is substituted in the Taylor expansion and the quadratic and higher-order terms are ignored, the Euler method arises. The Taylor expansion is used below to analyse the error committed by the Euler method, and it can be extended to produce Runge–Kutta methods.

A closely related derivation is to substitute the forward finite difference formula for the derivative,

$$y'(t_0) \approx \frac{y(t_0 + h) - y(t_0)}{h}$$

in the differential equation $y' = f(t, y)$. Again, this yields the Euler method. A similar computation leads to the midpoint rule and the backward Euler method. Finally, one can integrate the differential equation from t_0 to $t_0 + h$ and apply the fundamental theorem of calculus to get:

$$y(t_0 + h) - y(t_0) = \int_{t_0}^{t_0+h} f(t, y(t))\mathrm{d}t.$$

Now approximate the integral by the left-hand rectangle method (with only one rectangle):

$$\int_{t_0}^{t_0+h} f(t, y(t))\mathrm{d}t \approx hf(t_0, y(t_0)).$$

Combining both equations, one finds again the Euler method. This line of thought can be continued to arrive at various linear multistep methods.

Rounding Errors

The discussion up to now has ignored the consequences of rounding error. In step n of the Euler method, the rounding error is roughly of the magnitude εy_n where ε is the machine epsilon.

Assuming that the rounding errors are all of approximately the same size, the combined rounding error in N steps is roughly $N\varepsilon y_0$ if all errors points in the same direction. Since the number of steps is inversely proportional to the step size h, the total rounding error is proportional to ε / h.

In reality, however, it is extremely unlikely that all rounding errors point in the same direction. If instead it is assumed that the rounding errors are independent rounding variables, then the total rounding error is proportional to $\varepsilon / \sqrt{h}$.

Thus, for extremely small values of the step size, the truncation error will be small but the effect of rounding error may be big. Most of the effect of rounding error can be easily avoided if compensated summation is used in the formula for the Euler method.

Modifications and Extensions

A simple modification of the Euler method which eliminates the stability problems noted in the previous section is the backward Euler method:

$$y_{n+1} = y_n + hf(t_{n+1}, y_{n+1}).$$

This differs from the (standard, or forward) Euler method in that the function f is evaluated at the end point of the step, instead of the starting point.

The backward Euler method is an implicit method, meaning that the formula for the backward Euler method has y_{n+1} on both sides, so when applying the backward Euler method we have to solve an equation. This makes the implementation more costly.

Other modifications of the Euler method that help with stability yield the exponential Euler method or the semi-implicit Euler method.

More complicated methods can achieve a higher order (and more accuracy). One possibility is to use more function evaluations. This is illustrated by the midpoint method which is already mentioned in this article:

$$y_{n+1} = y_n + hf\left(t_n + \tfrac{1}{2}h, y_n + \tfrac{1}{2}hf(t_n, y_n)\right).$$

This leads to the family of Runge–Kutta methods. The other possibility is to use more past values, as illustrated by the two-step Adams–Bashforth method:

$$y_{n+1} = y_n + \tfrac{3}{2}hf(t_n, y_n) - \tfrac{1}{2}hf(t_{n-1}, y_{n-1}).$$

This leads to the family of linear multistep methods.

Finite Difference Method

In mathematics, finite-difference methods are numerical methods for approximating the solutions to differential equations using finite difference equations to approximate derivatives.

Derivation from Taylor's Polynomial

Assuming the function whose derivatives are to be approximated is properly-behaved, by Taylor's theorem,

$$f(x_0 + h) = f(x_0) + \frac{f'(x_0)}{1!}h + \frac{f^{(2)}(x_0)}{2!}h^2$$

$$+ \cdots + \frac{f^{(n)}(x_0)}{n!}h^n + R_n(x),$$

where $n!$ denotes the factorial of n, and $R_n(x)$ is a remainder term, denoting the difference between the Taylor polynomial of degree n and the original function. Again using the first derivative of the function f as an example, by Taylor's theorem,

$$f(x_0+h)=f(x_0)+f'(x_0)h+R_1(x),$$

Setting, x_0=a and (x-a)=h we have,

$$f(a+h)=f(a)+f'(a)h+R_1(x),$$

Dividing across by h gives:

$$\frac{f(a+h)}{h}=\frac{f(a)}{h}+f'(a)+\frac{R_1(x)}{h}$$

Solving for f'(a):

$$f'(a)=\frac{f(a+h)-f(a)}{h}-\frac{R_1(x)}{h}$$

so that for $R_1(x)$ sufficiently small,

$$f'(a)\approx\frac{f(a+h)-f(a)}{h}.$$

Example: Ordinary Differential Equation

For example, consider the ordinary differential equation

$$u'(x)=3u(x)+2.$$

The Euler method for solving this equation uses the finite difference quotient

$$\frac{u(x+h)-u(x)}{h}\approx u'(x)$$

to approximate the differential equation by first substituting in for u'(x) and applying a little algebra to get

$$u(x+h)=u(x)+h(3u(x)+2).$$

The last equation is a finite-difference equation, and solving this equation gives an approximate solution to the differential equation.

Example: The Heat Equation

Consider the normalised heat equation in one dimension, with homogeneous Dirichlet boundary conditions

$$U_t = U_{xx}$$

$$U(0,t) = U(1,t) = 0 \text{ (boundary condition)}$$

$$U(x,0) = U_0(x) \text{ (initial condition)}$$

One way to numerically solve this equation is to approximate all the derivatives by finite differences. We partition the domain in space using a mesh $x_0,...,x_J$ and in time using a mesh $t_0,....,t_N$. We assume a uniform partition both in space and in time, so the difference between two consecutive space points will be h and between two consecutive time points will be k. The points

$$u(x_j,t_n) = u_j^n$$

will represent the numerical approximation of $u(x_j,t_n)$.

Crank–Nicolson Method

In numerical analysis, the Crank–Nicolson method is a finite difference method used for numerically solving the heat equation and similar partial differential equations.

It is a second-order method in time, it is implicit in time and can be written as an implicit Runge–Kutta method, and it is numerically stable. The method was developed by John Crank and Phyllis Nicolson in the mid 20th century.

For diffusion equations (and many other equations), it can be shown the Crank–Nicolson method is unconditionally stable. However, the approximate solutions can still contain (decaying) spurious oscillations if the ratio of time step Δt times the thermal diffusivity to the square of space step, Δx^2, is large (typically larger than 1/2).

For this reason, whenever large time steps or high spatial resolution is necessary, the less accurate backward Euler method is often used, which is both stable and immune to oscillations.

The Method

The Crank–Nicolson method is based on central difference in space, and the trapezoidal rule in time, giving second-order convergence in time.

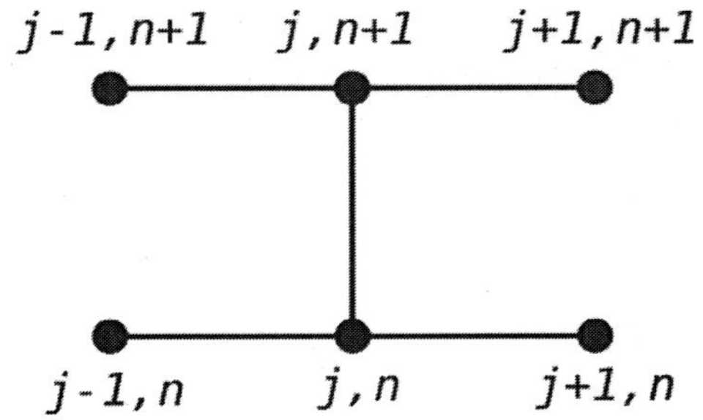

Figure: *The Crank–Nicolson stencil for a 1D problem.*

For example, in one dimension, if the partial differential equation is

$$\frac{\partial u}{\partial t} = F\left(u, x, t, \frac{\partial u}{\partial x}, \frac{\partial^2 u}{\partial x^2}\right)$$

then, letting $u(i\Delta x, n\Delta t) = u_i^n$, the equation for Crank–Nicolson method is a combination of the forward Euler method at n and the backward Euler method at $n + 1$ (note, however, that the method itself is *not* simply the average of those two methods, as the equation has an implicit dependence on the solution):

$$\frac{u_i^{n+1} - u_i^n}{\Delta t} = F_i^n\left(u, x, t, \frac{\partial u}{\partial x}, \frac{\partial^2 u}{\partial x^2}\right) \qquad \text{(forward Euler)}$$

$$\frac{u_i^{n+1} - u_i^n}{\Delta t} = F_i^{n+1}\left(u, x, t, \frac{\partial u}{\partial x}, \frac{\partial^2 u}{\partial x^2}\right) \qquad \text{(backward Euler)}$$

$$\frac{u_i^{n+1} - u_i^n}{\Delta t} = \frac{1}{2}\left[F_i^{n+1}\left(u, x, t, \frac{\partial u}{\partial x}, \frac{\partial^2 u}{\partial x^2}\right) + F_i^n\left(u, x, t, \frac{\partial u}{\partial x}, \frac{\partial^2 u}{\partial x^2}\right)\right]$$

(Crank-Nicolson)

The function F must be discretised spatially with a central difference.

Note that this is an *implicit method*: to get the "next" value of u in time, a system of algebraic equations must be solved. If the partial differential equation is nonlinear, the discretisation will also be nonlinear so that advancing in time will involve the solution of a system of nonlinear algebraic equations, though linearisations are possible.

In many problems, especially linear diffusion, the algebraic problem is tridiagonal and may be efficiently solved with the tridiagonal matrix algorithm, which gives a fast $\mathcal{O}(n)$ direct solution as opposed to the usual $\mathcal{O}(n^3)$ for a full matrix.

Example: 1D Diffusion

The Crank–Nicolson method is often applied to diffusion problems. As an example, for linear diffusion,

$$\frac{\partial u}{\partial t} = a\frac{\partial^2 u}{\partial x^2}$$

whose Crank–Nicolson discretisation is then:

$$\frac{u_i^{n+1} - u_i^n}{\Delta t} = \frac{a}{2(\Delta x)^2}\left((u_{i+1}^{n+1} - 2u_i^{n+1} + u_{i-1}^{n+1}) + (u_{i+1}^n - 2u_i^n + u_{i-1}^n)\right)$$

or, letting $r = \dfrac{a\Delta t}{2(\Delta x)^2}$:

$$-ru_{i+1}^{n+1} + (1+2r)u_i^{n+1} - ru_{i-1}^{n+1} = ru_{i+1}^n + (1-2r)u_i^n + ru_{i-1}^n$$

which is a tridiagonal problem, so that u_i^{n+1} may be efficiently solved by using the tridiagonal matrix algorithm in favour of a much more costly matrix inversion.

A quasilinear equation, such as (this is a minimalistic example and not general)

$$\frac{\partial u}{\partial t} = a(u)\frac{\partial^2 u}{\partial x^2}$$

would lead to a nonlinear system of algebraic equations which could not be easily solved as above; however, it is possible in some cases to linearise the problem by using the old value for a, that is $a_i^n(u)$ instead of $a_i^{n+1}(u)$.

Other times, it may be possible to estimate using an explicit method and maintain stability.

Example: 1D Diffusion with Advection for Steady Flow, with Multiple Channel Connections

This is a solution usually employed for many purposes when there's a contamination problem in streams or rivers under steady flow conditions but information is given in one dimension only.

Often the problem can be simplified into a 1-dimensional problem and still yield useful information.

Here we model the concentration of a solute contaminant in water. This problem is composed of three parts:

the known diffusion equation (D_x chosen as constant), an advective component (which means the system is evolving in space due to a velocity field), which we choose to be a constant Ux, and a lateral interaction between longitudinal channels (k).

$$\langle 0\rangle \frac{\partial C}{\partial t} = D_x \frac{\partial^2 C}{\partial x^2} - U_x \frac{\partial C}{\partial x} - k(C - C_N) - k(C - C_M)$$

where C is the concentration of the contaminant and subscripts N and M correspond to *previous* and *next* channel.

The Crank–Nicolson method (where i represents position and j time) transforms each component of the PDE into the following:

$$\langle 1\rangle \frac{\partial C}{\partial t} = \frac{C_i^{j+1} - C_i^j}{\Delta t}$$

$$\langle 2\rangle \frac{\partial^2 C}{\partial x^2} = \frac{1}{2(\Delta x)^2}\left((C_{i+1}^{j+1} - 2C_i^{j+1} + C_{i-1}^{j+1}) + (C_{i+1}^j - 2C_i^j + C_{i-1}^j)\right)$$

$$\langle 3\rangle \frac{\partial C}{\partial x} = \frac{1}{2}\left(\frac{(C_{i+1}^{j+1} - C_{i-1}^{j+1})}{2(\Delta x)} + \frac{(C_{i+1}^j - C_{i-1}^j)}{2(\Delta x)}\right)$$

$$\langle 4\rangle C = \frac{1}{2}(C_i^{j+1} + C_i^j)$$

$$\langle 5\rangle C_N = \frac{1}{2}(C_{Ni}^{j+1} + C_{Ni}^j)$$

$$\langle 6\rangle C_M = \frac{1}{2}(C_{Mi}^{j+1} + C_{Mi}^j)$$

Now we create the following constants to simplify the algebra:

$$\lambda = \frac{D_x \Delta t}{2\Delta x^2}$$

$$\alpha = \frac{U_x \Delta t}{4\Delta x}$$

$$\beta = \frac{k\Delta t}{2}$$

and substitute <1>, <2>, <3>, <4>, <5>, <6>, α, β and λ into <0>.

We then put the *new time* terms on the left (j + 1) and the *present time* terms on the right (j) to get:

$$-\beta C_{Ni}^{j+1} - (\lambda+\alpha)C_{i-1}^{j+1} + (1+2\lambda+2\beta)C_i^{j+1}$$
$$-(\lambda-\alpha)C_{i+1}^{j+1} - \beta C_{Mi}^{j+1} = \beta C_{Ni}^{j} + (\lambda+\alpha)C_{i-1}^{j} +$$
$$(1-2\lambda-2\beta)C_i^j + (\lambda-\alpha)C_{i+1}^j + \beta C_{Mi}^j$$

To model the *first* channel, we realise that it can only be in contact with the following channel (M), so the expression is simplified to:

$$-(\lambda+\alpha)C_{i-1}^{j+1} + (1+2\lambda+\beta)C_i^{j+1} - (\lambda-\alpha)C_{i+1}^{j+1}$$
$$-\beta C_{Mi}^{j+1} = +(\lambda+\alpha)C_{i-1}^{j} + (1-2\lambda-\beta)C_i^j + (\lambda-\alpha)C_{i+1}^j + \beta C_{Mi}^j$$

In the same way, to model the *last* channel, we realise that it can only be in contact with the previous channel (N), so the expression is simplified to:

$$-\beta C_{Ni}^{j+1} - (\lambda+\alpha)C_{i-1}^{j+1} + (1+2\lambda+\beta)C_i^{j+1}$$
$$-(\lambda-\alpha)C_{i+1}^{j+1} = \beta C_{Ni}^{j} + (\lambda+\alpha)C_{i-1}^{j}$$
$$+(1-2\lambda-\beta)C_i^j + (\lambda-\alpha)C_{i+1}^j$$

To solve this linear system of equations we must now see that boundary conditions must be given first to the beginning of the channels:

C_0^j: initial condition for the channel at present time step

C_0^{j+1} : initial condition for the channel at next time step

C_{N0}^j : initial condition for the previous channel to the one analysed at present time step

C_{M0}^j : initial condition for the next channel to the one analysed at present time step

For the last cell of the channels (z) the most convenient condition becomes an adiabatic one, so

$$\frac{\partial C}{\partial x}_{x=z} = \frac{(C_{i+1} - C_{i-1})}{2\Delta x} = 0$$

This condition is satisfied if and only if (regardless of a null value)

$$C_{i+1}^{j+1} = C_{i-1}^{j+1}$$

Let us solve this problem (in a matrix form) for the case of 3 channels and 5 nodes (including the initial boundary condition).

We express this as a linear system problem:

$$\left[AA\right]\left[C^{j+1}\right] = [BB][C^{j}] + [d]$$

where

$$\mathbf{C^{j+1}} = \begin{bmatrix} C_{11}^{j+1} \\ C_{12}^{j+1} \\ C_{13}^{j+1} \\ C_{14}^{j+1} \\ C_{21}^{j+1} \\ C_{22}^{j+1} \\ C_{23}^{j+1} \\ C_{24}^{j+1} \\ C_{31}^{j+1} \\ C_{32}^{j+1} \\ C_{33}^{j+1} \\ C_{34}^{j+1} \end{bmatrix} \quad \text{and} \quad \mathbf{C^{j}} = \begin{bmatrix} C_{11}^{j} \\ C_{12}^{j} \\ C_{13}^{j} \\ C_{14}^{j} \\ C_{21}^{j} \\ C_{22}^{j} \\ C_{23}^{j} \\ C_{24}^{j} \\ C_{31}^{j} \\ C_{32}^{j} \\ C_{33}^{j} \\ C_{34}^{j} \end{bmatrix}$$

Now we must realise that AA and BB should be arrays made of four different subarrays (remember that only three channels are considered for this example but it covers the main part discussed above).

$$\mathbf{AA} = \begin{bmatrix} AA1 & AA3 & 0 \\ AA3 & AA2 & AA3 \\ 0 & AA3 & AA1 \end{bmatrix} \text{ and}$$

$$\mathbf{BB} = \begin{bmatrix} BB1 & -AA3 & 0 \\ -AA3 & BB2 & -AA3 \\ 0 & -AA3 & BB1 \end{bmatrix}$$

where the elements mentioned above correspond to the next arrays and an additional 4x4 full of zeros. Please note that the sizes of AA and BB are 12x12:

$$\mathbf{AA1} = \begin{bmatrix} (1+2\lambda+\beta) & -(\lambda-\alpha) & 0 & 0 \\ -(\lambda+\alpha) & (1+2\lambda+\beta) & -(\lambda-\alpha) & 0 \\ 0 & -(\lambda+\alpha) & (1+2\lambda+\beta) & -(\lambda-\alpha) \\ 0 & 0 & -2\lambda & (1+2\lambda+\beta) \end{bmatrix},$$

$$\mathbf{AA2} = \begin{bmatrix} (1+2\lambda+2\beta) & -(\lambda-\alpha) & 0 & 0 \\ -(\lambda+\alpha) & (1+2\lambda+2\beta) & -(\lambda-\alpha) & 0 \\ 0 & -(\lambda+\alpha) & (1+2\lambda+2\beta) & -(\lambda-\alpha) \\ 0 & 0 & -2\lambda & (1+2\lambda+2\beta) \end{bmatrix},$$

$$\mathbf{AA3} = \begin{bmatrix} -\beta & 0 & 0 & 0 \\ 0 & -\beta & 0 & 0 \\ 0 & 0 & -\beta & 0 \\ 0 & 0 & 0 & -\beta \end{bmatrix},$$

$$\mathbf{BB1} = \begin{bmatrix} (1-2\lambda-\beta) & (\lambda-\alpha) & 0 & 0 \\ (\lambda+\alpha) & (1-2\lambda-\beta) & (\lambda-\alpha) & 0 \\ 0 & (\lambda+\alpha) & (1-2\lambda-\beta) & (\lambda-\alpha) \\ 0 & 0 & 2\lambda & (1-2\lambda-\beta) \end{bmatrix}$$

and

$$\mathbf{BB2} = \begin{bmatrix} (1-2\lambda-2\beta) & (\lambda-\alpha) & 0 & 0 \\ (\lambda+\alpha) & (1-2\lambda-2\beta) & (\lambda-\alpha) & 0 \\ 0 & (\lambda+\alpha) & (1-2\lambda-2\beta) & (\lambda-\alpha) \\ 0 & 0 & 2\lambda & (1-2\lambda-2\beta) \end{bmatrix}.$$

The d vector here is used to hold the boundary conditions. In this example it is a 12x1 vector:

$$\mathbf{d} = \begin{bmatrix} (\lambda+\alpha)(C_{10}^{j+1}+C_{10}^{j}) \\ 0 \\ 0 \\ 0 \\ (\lambda+\alpha)(C_{20}^{j+1}+C_{20}^{j}) \\ 0 \\ 0 \\ 0 \\ (\lambda+\alpha)(C_{30}^{j+1}+C_{30}^{j}) \\ 0 \\ 0 \\ 0 \end{bmatrix}$$

To find the concentration at any time, one must iterate the following equation:

$$\left[C^{j+1}\right]=\left[AA^{-1}\right]([BB][C^{j}]+[d])$$

Example: 2D Diffusion

When extending into two dimensions on a uniform Cartesian grid, the derivation is similar and the results may lead to a system of band-diagonal equations rather than tridiagonal ones.

The two-dimensional heat equation

$$\frac{\partial u}{\partial t}=a\left(\frac{\partial^2 u}{\partial x^2}+\frac{\partial^2 u}{\partial y^2}\right)$$

can be solved with the Crank–Nicolson discretisation of

$$\begin{aligned} u_{i,j}^{n+1} &= u_{i,j}^{n}+\frac{1}{2}\frac{a\Delta t}{(\Delta x)^2}\Big[(u_{i+1,j}^{n+1}+u_{i-1,j}^{n+1}+u_{i,j+1}^{n+1}+u_{i,j-1}^{n+1}-4u_{i,j}^{n+1}) \\ &\quad +(u_{i+1,j}^{n}+u_{i-1,j}^{n}+u_{i,j+1}^{n}+u_{i,j-1}^{n}-4u_{i,j}^{n})\Big] \end{aligned}$$

assuming that a square grid is used so that $\Delta x=\Delta y$. This equation can be simplified somewhat by rearranging terms and using the CFL number

$$\mu = \frac{a\Delta t}{(\Delta x)^2}$$

For the Crank–Nicolson numerical scheme, a low CFL number is not required for stability, however it is required for numerical accuracy. We can now write the scheme as:

$$(1+2\mu)u_{i,j}^{n+1} - \frac{\mu}{2}\left(u_{i+1,j}^{n+1} + u_{i-1,j}^{n+1} + u_{i,j+1}^{n+1} + u_{i,j-1}^{n+1}\right)$$
$$= (1-2\mu)u_{i,j}^{n} + \frac{\mu}{2}\left(u_{i+1,j}^{n} + u_{i-1,j}^{n} + u_{i,j+1}^{n} + u_{i,j-1}^{n}\right)$$

Application in Financial Mathematics

Because a number of other phenomena can be modelled with the heat equation (often called the diffusion equation in financial mathematics), the Crank–Nicolson method has been applied to those areas as well. Particularly, the Black–Scholes option pricing model's differential equation can be transformed into the heat equation, and thus numerical solutions for option pricing can be obtained with the Crank–Nicolson method.

The importance of this for finance, is that option pricing problems, when extended beyond the standard assumptions (e.g. incorporating changing dividends), cannot be solved in closed form, but can be solved using this method. Note however, that for non-smooth final conditions (which happen for most financial instruments), the Crank–Nicolson method is not satisfactory as numerical oscillations are not damped. For vanilla options, this results in oscillation in the gamma value around the strike price. Therefore, special damping initialisation steps are necessary (e.g., fully implicit finite difference method).

Chapter 11

Lyapunov Stability

Various types of stability may be discussed for the solutions of differential equations describing dynamical systems. The most important type is that concerning the stability of solutions near to a point of equilibrium. This may be discussed by the theory of Lyapunov. In simple terms, if all solutions of the dynamical system that start out near an equilibrium point x_e stay near forever, then is Lyapunov stable. More strongly, if is Lyapunov stable and all solutions that start out near converge to, then is asymptotically stable.

The notion of exponential stability guarantees a minimal rate of decay, i.e., an estimate of how quickly the solutions converge. The idea of Lyapunov stability can be extended to infinite-dimensional manifolds, where it is known as structural stability, which concerns the behaviour of different but "nearby" solutions to differential equations. Input-to-state stability (ISS) applies Lyapunov notions to systems with inputs.

History

Lyapunov stability is named after Aleksandr Lyapunov, a Russian mathematician who published his book "The General Problem of Stability of Motion" in 1892. Lyapunov was the first to consider the modifications necessary in nonlinear systems to the linear theory of stability based on linearising near a point of equilibrium. His work, initially published in Russian and then translated to French, received little attention for many years. Interest in it started suddenly during the Cold War (1953-1962) period when the so-called "Second Method of Lyapunov" was found to be applicable to the stability of aerospace guidance systems which typically contain strong nonlinearities not treatable by other methods.

A large number of publications appeared then and since in the control and systems literature. More recently the concept of the Lyapunov exponent (related to Lyapunov's First Method of discussing stability) has received wide interest in connection with chaos theory. Lyapunov stability methods have also been applied to finding equilibrium solutions in traffic assignment problems.

Lyapunov's Second Method for Stability

Lyapunov, in his original 1892 work, proposed two methods for demonstrating stability. The first method developed the solution in a series which was then proved convergent within limits. The second method, which is almost universally used nowadays, makes use of a *Lyapunov function V(x)* which has an analogy to the potential function of classical dynamics. It is introduced as follows for a system having a point of equilibrium at x = 0. Consider a function $V(x):\mathbb{R}^n \to \mathbb{R}$ such that

- $V(x) \geq 0$ with equality if and only if $x = 0$ (positive definite)
- $\dot{V}(x) = \frac{d}{dt}V(x) \leq 0$ with equality if and only if $x = 0$ (negative definite).

Then *V(x)* is called a Lyapunov function candidate and the system is asymptotically stable in the sense of Lyapunov (i.s.L.). (Note that $V(0) = 0$ is required; otherwise for example $V(x) = 1/(1+|x|)$ would "prove" that $\dot{x}(t) = x$ is locally stable. An additional condition called "properness" or "radial unboundedness" is required in order to conclude global asymptotic stability.)

It is easier to visualise this method of analysis by thinking of a physical system (e.g. vibrating spring and mass) and considering the energy of such a system.

If the system loses energy over time and the energy is never restored then eventually the system must grind to a stop and reach some final resting state. This final state is called the attractor.

However, finding a function that gives the precise energy of a physical system can be difficult, and for abstract mathematical systems, economic systems or biological systems, the concept of energy may not be applicable.

Lyapunov's realisation was that stability can be proven without requiring knowledge of the true physical energy, provided a Lyapunov function can be found to satisfy the above constraints.

Definition for Discrete-time Systems

The definition for discrete-time systems is almost identical to that for continuous-time systems. The definition below provides this, using an alternate language commonly used in more mathematical texts.

Let (X,d) be a metric space and $f: X \to X$ a continuous function. A point $x \in X$ is said to be Lyapunov stable, if, for each $\epsilon > 0$, there is a $\delta > 0$ such that for all $y \in X$, if

$$d(x,y) < \delta$$

then

$$d(f^n(x), f^n(y)) < \epsilon$$

for all $n \in \mathbb{N}$.

We say that x is asymptotically stable if it belongs to the interior of its stable set, *i.e.* if there is a such that

$$\lim_{n \to \infty} d(f^n(x), f^n(y)) = 0$$

whenever $d(x,y) < \delta$.

Stability for Linear State Space Models

A linear state space model

$$\dot{\mathbf{x}} = A\mathbf{x}$$

is asymptotically stable (in fact, exponentially stable) if all real parts of the eigenvalues of A are negative. This condition is equivalent to the following one:

$$A^T M + MA + N = 0$$

has a solution where $N = N^T > 0$ and $M = M^T > 0$ (positive definite matrices). (The relevant Lyapunov function is $V(x) = x^T Mx$.)

Correspondingly, a time-discrete linear state space model

$$\mathbf{x}_{t+1} = A\mathbf{x}_t$$

is asymptotically stable (in fact, exponentially stable) if all the eigenvalues of A have a modulus smaller than one.

This latter condition has been generalised to switched systems: a linear switched discrete time system (ruled by a set of matrices $\{A_1,\ldots,A_m\}$)

$$\mathbf{x}_{t+1} = A_{i_t}\mathbf{x}_t, \quad A_{i_t} \in \{A_1,\ldots,A_m\}$$

is asymptotically stable (in fact, exponentially stable) if the joint spectral radius of the set is smaller than one.

Stability for Systems with Inputs

A system with inputs (or controls) has the form

$$\dot{\mathbf{x}} = f(\mathbf{x},\mathbf{u})$$

where the (generally time-dependent) input u(t) may be viewed as a *control*, *external input*, *stimulus*, *disturbance*, or *forcing function*. The study of such systems is the subject of control theory and applied in control engineering.

For systems with inputs, one must quantify the effect of inputs on the stability of the system. The main two approaches to this analysis are BIBO stability (for linear systems) and input-to-state (ISS) stability (for nonlinear systems)

Example

Consider an equation, where compared to the Van der Pol oscillator equation the friction term is changed:

$$\ddot{y} + y - \varepsilon\left(\frac{\dot{y}^3}{3} - \dot{y}\right) = 0.$$

The equilibrium is at : $\ddot{y} = y = 0$. Here is a good example of an unsuccessful try to find a Lyapunov function that proves stability:

Let

$$x_1 = y, x_2 = \dot{y}$$

so that the corresponding system is

$$\dot{x}_2 = -x_1 + \varepsilon\left(\frac{x_2^{\,3}}{3} - x_2\right).$$

Let us choose as a Lyapunov function

$$V = \frac{1}{2}\left(x_1^2 + x_2^2\right)$$

which is clearly positive definite. Its derivative is

$$\begin{aligned}\dot{V} &= x_1\dot{x}_1 + x_2\dot{x}_2 \\ &= x_1x_2 - x_1x_2 + \varepsilon\left(\frac{x_2^4}{3} - x_2^2\right) \\ &= -\varepsilon\left(x_2^2 - \frac{x_2^4}{3}\right).\end{aligned}$$

It seems that if the parameter ε is positive, stability is asymptotic for $x_2^2 < 3$. But this is wrong, since $\dot{V}$ does not depend on x_1, and will be 0 everywhere on the axis.

Exponential Stability

In control theory, a continuous linear time-invariant system is exponentially stable if and only if the system has eigenvalues (i.e., the poles of input-to-output systems) with strictly negative real parts. (i.e., in the left half of the complex plane). A discrete-time input-to-output LTI system is exponentially stable if and only if the poles of its transfer function lie strictly within the unit circle centred on the origin of the complex plane.

Exponential stability is a form of asymptotic stability. Systems that are not LTI are exponentially stable if their convergence is bounded by exponential decay.

Practical Consequences

An exponentially stable LTI system is one that will not "blow up" (i.e., give an unbounded output) when given a finite input or non-zero initial condition. Moreover, if the system is given a fixed, finite input (i.e., a step), then any resulting oscillations in the output will decay at an exponential rate, and the output will tend asymptotically to a new final, steady-state value. If the system is instead given a Dirac delta impulse as input, then induced oscillations will die away and the system will return to its previous value. If oscillations do not die away, or the system does not return to its original output when an impulse is applied, the system is instead marginally stable.

Example Exponentially Stable LTI Systems

The graph on the right shows the impulse response of two similar systems. The green curve is the response of the system with impulse response $y(t) = e^{-\frac{t}{5}}$, while the blue represents the system $y(t) = e^{-\frac{t}{5}} \sin(t)$. Although one response is oscillatory, both return to the original value of 0 over time.

Real-world Example

Imagine putting a marble in a ladle. It will settle itself into the lowest point of the ladle and, unless disturbed, will stay there. Now imagine giving the ball a push, which is an approximation to a Dirac delta impulse.

The marble will roll back and forth but eventually resettle in the bottom of the ladle. Drawing the horizontal position of the marble over time would give a gradually diminishing sinusoid rather like the blue curve in the image above. A step input in this case requires supporting the marble away from the bottom of the ladle, so that it cannot roll back. It will stay in the same position and will not, as would be the case if the system were only marginally stable or entirely unstable, continue to move away from the bottom of the ladle under this constant force equal to its weight.

It is important to note that in this example the system is not stable for all inputs. Give the marble a big enough push, and it will fall out of the ladle and fall, stopping only when it reaches the floor. For some systems, therefore, it is proper to state that a system is exponentially stable *over a certain range of inputs*.

Rate of Convergence

In numerical analysis, the speed at which a convergent sequence approaches its limit is called the rate of convergence. Although strictly speaking, a limit does not give information about any finite first part of the sequence, this concept is of practical importance if we deal with a sequence of successive approximations for an iterative method, as then typically fewer iterations are needed to yield a useful approximation if the rate of convergence is higher.

This may even make the difference between needing ten or a million iterations. Similar concepts are used for discretisation methods. The solution of the discretised problem converges to the solution of the continuous problem as the grid size goes to zero, and the speed of convergence is one of the factors of the efficiency of the method.

However, the terminology in this case is different from the terminology for iterative methods. Series acceleration is a collection of techniques for improving the rate of convergence of a series discretisation. Such acceleration is commonly accomplished with sequence transformations.

Convergence Speed for Iterative Methods

Basic Definition: Suppose that the sequence $\{x_k\}$ converges to the number L. We say that this sequence converges linearly to L, if there exists a number μ – (0, 1) such that

$$\lim_{k\to\infty}\frac{|x_{k+1}-L|}{|x_k-L|}=\mu.$$

The number μ is called the *rate of convergence.*

If the sequences converges, and

- μ = 0, then the sequence is said to converge superlinearly.
- μ = 1, then the sequence is said to converge sublinearly.

If the sequences *converges sublinearly* and additionally

$$\lim_{k\to\infty}\frac{|x_{k+2}-x_{k+1}|}{|x_{k+1}-x_k|}=1,$$

then it is said the sequence $\{x_k\}$ converges logarithmically to L.

The next definition is used to distinguish superlinear rates of convergence. We say that the sequence converges with order q to L for q>1 if

$$\lim_{k\to\infty}\frac{|x_{k+1}-L|}{|x_k-L|^q}=\mu| \mu>0.$$

In particular, convergence with order

- 2 is called quadratic convergence,
- 3 is called cubic convergence,
- etc.

This is sometimes called *Q-linear convergence*, *Q-quadratic convergence*, etc., to distinguish it from the definition below. The Q stands for "quotient," because the definition uses the quotient between two successive terms.

Extended Definition: The drawback of the above definitions is that these do not catch some sequences which still converge reasonably fast, but whose "speed" is variable, such as the sequence $\{b_k\}$ below. Therefore, the definition of rate of convergence is sometimes extended as follows.

Under the new definition, the sequence $\{x_k\}$ converges with at least order q if there exists a sequence $\{\varepsilon_k\}$ such that

$$|x_k - L| \le \varepsilon_k^q \quad \text{for all } k,$$

and the sequence $\{\varepsilon_k\}$ converges to zero with order q according to the above "simple" definition. To distinguish it from that definition, this is sometimes called *R-linear convergence*, *R-quadratic convergence*, etc. (with the R standing for "root").

Convergence Speed for Discretisation Methods

A similar situation exists for discretisation methods. The important parameter here for the convergence speed is not the iteration number k but it depends on the number of grid points and grid spacing. In this case, the number of grid points in a discretisation process is inversely proportional to the grid spacing here it denoted as n. In this case, a sequence x_n is said to converge to L with order p if there exists a constant C such that

$$|x_n - L| < Cn^{-p} \text{ for all } n.$$

This is written as $|x_n - L| = O(n^{-p})$ using the big O notation. This is the relevant definition when discussing methods for numerical quadrature or the solution of ordinary differential equations.

Examples

The sequence $\{d_k\}$ with $d_k = 1 / (k+1)$ was introduced above. This sequence converges with order 1 according to the convention for discretisation methods. The sequence $\{a_k\}$ with $a_k = 2^{-k}$, which was also introduced above, converges with order p for every number p. It is said to converge exponentially using the convention for discretisation methods. However, it only converges linearly (that is, with order 1) using the convention for iterative methods.

Acceleration of Convergence

Many methods exist to increase the rate of convergence of a given sequence, i.e. to transform a given sequence into one converging faster to the same limit. Such techniques are in general known as "series acceleration". The goal of the transformed sequence is to be much less "expensive" to calculate than the original sequence. One example of series acceleration is Aitken's delta-squared process.

Power Series Solution of Differential Equations

In mathematics, the power series method is used to seek a power series solution to certain differential equations. In general, such a solution assumes a power series with unknown coefficients, then substitutes that solution into the differential equation to find a recurrence relation for the coefficients.

Method

Consider the second-order linear differential equation

$$a_2(z)f''(z) + a_1(z)f'(z) + a_0(z)f(z) = 0.$$

Suppose a_2 is nonzero for all z. Then we can divide throughout to obtain

$$f'' + \frac{a_1(z)}{a_2(z)}f' + \frac{a_0(z)}{a_2(z)}f = 0.$$

Suppose further that a_1/a_2 and a_0/a_2 are analytic functions. The power series method calls for the construction of a power series solution

$$f = \sum_{k=0}^{\infty} A_k z^k.$$

If a_2 is zero for some z, then the Frobenius method, a variation on this method, is suited to deal with so called *singular points*. The method works analogously for higher order equations as well as for systems.

Chapter 12

Homogeneous Functions

One is that a first-order ordinary differential equation is homogeneous (of degree 0) if it has the form

$$\frac{dy}{dx} = F(x, y)$$

where x is an independent variable, y a dependent variable, and $F(x, y)$ is a homogeneous function of degree n; in other words scalar multiplying each variable by a constant λ leaves the function unchanged:

$$F(\lambda x, \lambda y) = \lambda^n F(x, y).$$

In a more general form:

$$N(x, y)dx + M(x, y)dy = 0$$

where M and N are also homogeneous.

Solution Method

Since $F(\lambda\, x, \lambda\, y) = F(x, y)$ for all λ, so λ can be arbitrarily chosen to simplify the form of the equation.

One can solve this equation by making a simple change of variables $y = ux$ and then using the product rule on the left hand side as follows,

$$\frac{d(ux)}{dx} = x\frac{du}{dx} + u\frac{dx}{dx} = x\frac{du}{dx} + u$$

and then using the identity $F(\lambda\, x, \lambda y) = F(x, y)$ to simplify the right hand side by choosing to set $\lambda = 1/x$, transforming the original problem into the separable differential equation

$$x\frac{du}{dx}+u=F(1,u)$$

which can then be integrated by the usual methods.

Homogeneous Linear Equations

Another is that a linear differential equation can be represented as a linear operator acting on *y*. In this case a linear homogeneous differential equations is of the form

$$Ly=0$$

where *L* is a differential operator, a sum of derivatives, each multiplied by functions of *x*:

$$L=\sum_{i=1}^{n} f_i(x)\frac{d^i}{dx^i}=0.$$

Exact Differential Equation

In mathematics, an exact differential equation or total differential equation is a certain kind of ordinary differential equation which is widely used in physics and engineering.

Definition

Given a simply connected and open subset *D* of R^2 and two functions *I* and *J* which are continuous on *D* then an implicit first-order ordinary differential equation of the form

$$I(x,y)\mathrm{d}x+J(x,y)\mathrm{d}y=0,$$

is called exact differential equation if there exists a continuously differentiable function *F*, called the potential function, so that

$$\frac{\partial F}{\partial x}(x,y)=I$$

and

$$\frac{\partial F}{\partial y}(x,y)=J.$$

The nomenclature of "exact differential equation" refers to the exact derivative of a function. For a function $F(x_0,x_1,\ldots,x_{n-1},x_n)$, the exact or total derivative with respect to x_0 is given by

$$\frac{\mathrm{d}F}{\mathrm{d}x_0}=\frac{\partial F}{\partial x_0}+\sum_{i=1}^{n}\frac{\partial F}{\partial x_i}\frac{\mathrm{d}x_i}{\mathrm{d}x_0}.$$

Example

The function

$$F(x,y) := \frac{1}{2}(x^2 + y^2)$$

is a potential function for the differential equation

$$xx' + yy' = 0.$$

Existence of Potential Functions

In physical applications the functions I and J are usually not only continuous but even continuously differentiable. Schwarz's Theorem then provides us with a necessary criterion for the existence of a potential function.

For differential equations defined on simply connected sets the criterion is even sufficient and we get the following theorem:

Given a differential equation of the form (for example, when F has zero slope in the x and y direction at F(x, y)):

$$I(x,y)dx + J(x,y)dy = 0,$$

with I and J continuously differentiable on a simply connected and open subset D of R^2 then a potential function F exists if and only if

$$\frac{\partial I}{\partial y}(x,y) = \frac{\partial J}{\partial x}(x,y).$$

Solutions to Exact Differential Equations

Given an exact differential equation defined on some simply connected and open subset D of R^2 with potential function F then a differentiable function f with (x, $f(x)$) in D is a solution if and only if there exists real number c so that

$$F(x, f(x)) = c.$$

For an initial value problem

$$y(x_0) = y_0$$

we can locally find a potential function by

$$F(x,y) = \int_{x_0}^{x} I(t,y_0)dt + \int_{y_0}^{y} J(x,t)dt.$$

Solving

$$F(x,y)=c$$

for y, where c is a real number, we can then construct all solutions.

Complex Differential Equation

A complex differential equation is a differential equation whose solutions are functions of a complex variable. Constructing integrals involves choice of what path to take, which means singularities and branch points of the equation need to be studied.

Analytic continuation is used to generate new solutions and this means topological considerations such as monodromy, coverings and connectedness are to be taken into account.

Existence and uniqueness theorems involve the use of majorants and minorants. Study of rational second order ODEs in the complex plane led to the discovery of new transcendental special functions which are now known as Painlevé transcendents.

Nevanlinna theory can be used to study complex differential equations. This leads to extensions of Malmquist's theorem.

Generalisations

Generalisations include partial differential equations in several complex variables, or differential equations on complex manifolds.

Also there are at least a couple of ways of studying complex difference equations: either study holomorphic functions which satisfy functional relations given by the difference equation or study discrete analogs of holomorphicity such as monodiffric functions. Also integral equations can be studied in the complex domain.

History

Some of the early contributors to the theory of complex differential equations include: Pierre Boutroux, Paul Painlevé, Lazarus Fuchs, Henri Poincaré, David Hilbert, George David Birkhoff, Kosaku Yosida, Hans Wittich, Charles Briot, Jean Claude Bouquet, Johannes Malmquist.

Relation to Processes

Recurrence Relation

In mathematics, a recurrence relation is an equation that recursively defines a sequence, once one or more initial terms are given: each further term of the sequence is defined as a function of the preceding terms. The term difference equation sometimes (and for the purposes of this article) refers to a specific type of recurrence relation. However, "difference equation" is frequently used to refer to *any* recurrence relation.

An example of a recurrence relation is the logistic map:

$$x_{n+1} = rx_n(1-x_n),$$

with a given constant r; given the initial term x_0 each subsequent term is determined by this relation.

Some simply defined recurrence relations can have very complex (chaotic) behaviours, and they are a part of the field of mathematics known as nonlinear analysis. Solving a recurrence relation means obtaining a closed-form solution: a non-recursive function of n.

Fibonacci Numbers

The Fibonacci numbers are the archetype of a linear, homogeneous recurrence relation with constant coefficients. They are defined using the linear recurrence relation

$$F_n = F_{n-1} + F_{n-2}$$

with seed values:

$$F_0 = 0$$

$$F_1 = 1$$

Explicitly, recurrence yields the equations:

$$F_2 = F_1 + F_0$$

$$F_3 = F_2 + F_1$$

$$F_4 = F_3 + F_2$$

etc.

We obtain the sequence of Fibonacci numbers which begins:

0, 1, 1, 2, 3, 5, 8, 13, 21, 34, 55, 89, ...

It can be solved by methods described below yielding the closed form expression which involve powers of the two roots of the characteristic polynomial $t^2 = t + 1$; the generating function of the sequence is the rational function

$$\frac{t}{1-t-t^2}.$$

Structure

Linear Homogeneous Recurrence Relations with Constant Coefficients: An order d linear homogeneous recurrence relation with constant coefficients is an equation of the form

$$a_n = c_1 a_{n-1} + c_2 a_{n-2} + \cdots + c_d a_{n-d},$$

where the d coefficients c_i (for all i) are constants.

More precisely, this is an infinite list of simultaneous linear equations, one for each $n>d-1$. A sequence which satisfies a relation of this form is called a linear recurrence sequence or LRS. There are d degrees of freedom for LRS, i.e., the initial values $a_0,\ldots,a_{d-1}$ can be taken to be any values but then the linear recurrence determines the sequence uniquely.

The same coefficients yield the characteristic polynomial (also "auxiliary polynomial")

$$p(t) = t^d - c_1 t^{d-1} - c_2 t^{d-2} - \cdots - c_d$$

whose d roots play a crucial role in finding and understanding the sequences satisfying the recurrence. If the roots r_1, r_2, ... are all distinct, then the solution to the recurrence takes the form

$$a_n = k_1 r_1^n + k_2 r_2^n + \cdots + k_d r_d^n,$$

where the coefficients k_i are determined in order to fit the initial conditions of the recurrence. When the same roots occur multiple times, the terms in this formula corresponding to the second and later occurrences of the same root are multiplied by increasing powers of n. For instance, if the characteristic polynomial can be factored as $(x - r)^3$, with the same root r occurring three times, then the solution would take the form

$$a_n = k_1 r^n + k_2 n r^n + k_3 n^2 r^n.$$

Relationship to Difference Equations Narrowly Defined: Given an ordered sequence $\{a_n\}_{n=1}^{\infty}$ of real numbers: the first difference $\Delta(a_n)$ is defined as

$$\Delta(a_n) = a_{n+1} - a_n .$$

The second difference $\Delta^2(a_n)$ is defined as

$$\Delta^2(a_n) = \Delta(a_{n+1}) - \Delta(a_n)$$

which can be simplified to

$$\Delta^2(a_n) = a_{n+2} - 2a_{n+1} + a_n$$

More generally: the kth difference of the sequence a_n is written as $\Delta^k(a_n)$ is defined recursively as

$$\Delta^k(a_n) = \Delta^{k-1}(a_{n+1}) - \Delta^{k-1}(a_n) = \sum_{t=0}^{k} \binom{k}{t} (-1)^t a_{n+k-t}$$

(The sequence and its differences are related by a binomial transform.) The more restrictive definition of difference equation is an equation composed of a_n and its kth differences. (A widely used broader definition treats "difference equation" as synonymous with "recurrence relation".

Linear recurrence relations are difference equations, and conversely; since this is a simple and common form of recurrence, some authors use the two terms interchangeably. For example, the difference equation

$$3\Delta^2(a_n) + 2\Delta(a_n) + 7a_n = 0$$

is equivalent to the recurrence relation

$$3a_{n+2} = 4a_{n+1} - 8a_n$$

Thus one can solve many recurrence relations by rephrasing them as difference equations, and then solving the difference equation, analogously to how one solves ordinary differential equations. However, the Ackermann numbers are an example of a recurrence relation that do not map to a difference equation, much less points on the solution to a differential equation.

Summation equations relate to difference equations as integral equations relate to differential equations.

From Sequences to Grids: Single-variable or one-dimensional recurrence relations are about sequences (i.e. functions defined on one-dimensional grids). Multi-variable or n-dimensional recurrence relations are about n-dimensional grids. Functions defined on n-grids can also be studied with partial difference equations.

Solving

General Methods: For order 1, the recurrence

$$a_n = ra_{n-1}$$

has the solution $a_n = r^n$ with $a_0 = 1$ and the most general solution is $a_n = kr^n$ with $a_0 = k$. The characteristic polynomial equated to zero (the characteristic equation) is simply $t - r = 0$. Solutions to such recurrence relations of higher order are found by systematic means, often using the fact that $a_n = r^n$ is a solution for the recurrence exactly when $t = r$ is a root of the characteristic polynomial. This can be approached directly or using generating functions (formal power series) or matrices.

Consider, for example, a recurrence relation of the form

$$a_n = Aa_{n-1} + Ba_{n-2}.$$

When does it have a solution of the same general form as $a_n = r^n$? Substituting this guess (ansatz) in the recurrence relation, we find that

$$r^n = Ar^{n-1} + Br^{n-2} \text{ must be true for all } n > 1.$$

Dividing through by r^{n-2}, we get that all these equations reduce to the same thing:

$$r^2 = Ar + B,$$

$$r^2 - Ar - B = 0,$$

which is the characteristic equation of the recurrence relation. Solve for r to obtain the two roots λ_1, λ_2: these roots are known as the characteristic roots or eigenvalues of the characteristic equation. Different solutions are obtained depending on the nature of the roots: If these roots are distinct, we have the general solution

$$a_n = C\lambda_1^n + D\lambda_2^n$$

while if they are identical (when $A^2 + 4B = 0$), we have

$$a_n = C\lambda^n + Dn\lambda^n .$$

This is the most general solution; the two constants C and D can be chosen based on two given initial conditions a_0 and a_1 to produce a specific solution.

In the case of complex eigenvalues (which also gives rise to complex values for the solution parameters C and D), the use of complex numbers can be eliminated by rewriting the solution in trigonometric form.

In this case we can write the eigenvalues as $\lambda_1, \lambda_2 = \alpha \pm \beta i$. Then it can be shown that $a_n = C\lambda_1^n + D\lambda_2^n$ can be rewritten as

$$a_n = 2M^n\left(E\cos(\theta n) + F\sin(\theta n)\right) = 2GM^n\cos(\theta n - \delta),$$

where

$$M = \sqrt{\alpha^2 + \beta^2} \quad \cos\theta = \tfrac{\alpha}{M} \quad \sin\theta = \tfrac{\beta}{M}$$
$$C, D = E \mp Fi$$
$$G = \sqrt{E^2 + F^2} \quad \cos\delta = \tfrac{E}{G} \quad \sin\delta = \tfrac{F}{G}$$

Here E and F (or equivalently, G and δ) are real constants which depend on the initial conditions. In all cases—real distinct eigenvalues, real duplicated eigenvalues, and complex conjugate eigenvalues—the equation is stable (that is, the variable a converges to a fixed value (specifically, zero)); if and only if *both* eigenvalues are smaller than one in absolute value. In this second-order case, this condition on the eigenvalues can be shown to be equivalent to $|A| < 1 - B < 2$, which is equivalent to $|B| < 1$ and $|A| < 1 - B$.

The equation in the above example was homogeneous, in that there was no constant term. If one starts with the non-homogeneous recurrence

$$b_n = Ab_{n-1} + Bb_{n-2} + K$$

with constant term K, this can be converted into homogeneous form as follows: The steady state is found by setting $b_n = b_{n-1} = b_{n-2} = b^*$ to obtain

$$b^* = \frac{K}{1 - A - B}.$$

Then the non-homogeneous recurrence can be rewritten in homogeneous form as

$$[b_n - b^*] = A[b_{n-1} - b^*] + B[b_{n-2} - b^*],$$

which can be solved as above. The stability condition stated above in terms of eigenvalues for the second-order case remains valid for the general n^{th}-order case: the equation is stable if and only if all eigenvalues of the characteristic equation are less than one in absolute value.

Solving via Linear Algebra: Given a linearly recursive sequence, let C be the transpose of the companion matrix of its characteristic polynomial, that is

$$\begin{bmatrix} 0 & 1 & 0 & \cdots & 0 \\ 0 & 0 & 1 & \cdots & 0 \\ \vdots & \vdots & \vdots & \ddots & \vdots \\ 0 & 0 & 0 & \cdots & 1 \\ -c_0 & -c_1 & -c_2 & \cdots & -c_{d-1} \end{bmatrix}$$

where $T_n + c_{d-1}T_{n-1} + c_{d-2}T_{n-2} + \cdots + c_0 T_{n-d} = 0$ Call this matrix C. Observe that

$$\begin{bmatrix} a_n \\ \vdots \\ a_{n+(d-1)} \end{bmatrix} = C^n \begin{bmatrix} a_0 \\ \vdots \\ a_{d-1} \end{bmatrix}$$

Determine an eigenbasis $v_1, \ldots, v_d$ corresponding to eigenvalues $\lambda_1, \ldots, \lambda_d$. Then express the seed (the initial conditions of the LRS) as a linear combination of the eigenbasis vectors:

$$\begin{bmatrix} a_0 \\ \vdots \\ a_{d-1} \end{bmatrix} = b_1 v_1 + \cdots + b_d v_d$$

Then it conveniently works out that:

$$\begin{bmatrix} a_n \\ \vdots \\ a_{n+(d-1)} \end{bmatrix} = C^n \begin{bmatrix} a_0 \\ \vdots \\ a_{d-1} \end{bmatrix} = C^n (b_1 v_1 + \cdots + b_d v_d)$$

$$= \lambda_1^n b_1 v_1 + \cdots + \lambda_d^n b_d v_d$$

This description is really no different from general method above, however it is more succinct. It also works nicely for situations like

$$a_n = a_{n-1} - b_{n-1}.$$

$$b_n = 2a_{n-1} + b_{n-1}.$$

Where there are several linked recurrences.

Solving with z-transforms: Certain difference equations, in particular Linear constant coefficient difference equations, can be solved using z-transforms. The z-transforms are a class of integral transforms that lead to more convenient algebraic manipulations and more straightforward solutions. There are cases in which obtaining a direct solution would be all but impossible, yet solving the problem via a thoughtfully chosen integral transform is straightforward.

General Linear Homogeneous Recurrence Relations: Many linear homogeneous recurrence relations may be solved by means of the generalised hypergeometric series. Special cases of these lead to recurrence relations for the orthogonal polynomials, and many special functions. For example, the solution to

$$J_{n+1} = \frac{2n}{z} J_n - J_{n-1}$$

is given by

$$J_n = J_n(z),$$

the Bessel function, while

$$(b-n)M_{n-1} + (2n-b-z)M_n - nM_{n+1} = 0$$

is solved by

$$M_n = M(n,b;z)$$

the confluent hypergeometric series.

Solving a First Order Rational Difference Equation: A first order rational difference equation has the form $w_{t+1} = \frac{aw_t + b}{cw_t + d}$. Such an equation can be solved by writing w_t as a nonlinear transformation of another variable x_t which itself evolves linearly. Then standard methods can be used to solve the linear difference equation in.

Stability

Stability of Linear Higher-order Recurrences

The linear recurrence of order d,

$$a_n = c_1 a_{n-1} + c_2 a_{n-2} + \ldots + c_d a_{n-d},$$

has the characteristic equation

$$\lambda^d - c_1\lambda^{d-1} - c_2\lambda^{d-2} - \ldots - c_d\lambda^0 = 0.$$

The recurrence is stable, meaning that the iterates converge asymptotically to a fixed value, if and only if the eigenvalues (i.e., the roots of the characteristic equation), whether real or complex, are all less than unity in absolute value.

Stability of Linear First-order Matrix Recurrences

In the first-order matrix difference equation

$$[x_t - x^*] = A[x_{t-1} - x^*]$$

with state vector x and transition matrix A, x converges asymptotically to the steady state vector x^* if and only if all eigenvalues of the transition matrix A (whether real or complex) have an absolute value which is less than 1.

Stability of Nonlinear First-order Recurrences

Consider the nonlinear first-order recurrence

$$x_n = f(x_{n-1}).$$

This recurrence is locally stable, meaning that it converges to a fixed point x^* from points sufficiently close to x^*, if and only if the slope of f in the neighbourhood of x^* is smaller than unity in absolute value: that is,

$$| f'(x^*) | < 1.$$

A nonlinear recurrence could have multiple fixed points, in which case some fixed points may be locally stable and others locally unstable; for continuous f two adjacent fixed points cannot both be locally stable.

A nonlinear recurrence relation could also have a cycle of period k for $k > 1$. Such a cycle is stable, meaning that it attracts a set of initial conditions of positive measure, if the composite function $g(x) := f \circ f \circ \cdots \circ f(x)$ with f appearing k times is locally stable according to the same criterion:

$$|g'(x^*)|<1,$$

where x^* is any point on the cycle.

In a chaotic recurrence relation, the variable x stays in a bounded region but never converges to a fixed point or an attracting cycle; any fixed points or cycles of the equation are unstable.

Relationship to Differential Equations

When solving an ordinary differential equation numerically, one typically encounters a recurrence relation. For example, when solving the initial value problem

$$y'(t) = f(t, y(t)), \quad y(t_0) = y_0,$$

with Euler's method and a step size h, one calculates the values

$$y_0 = y(t_0), \quad y_1 = y(t_0 + h), \quad y_2 = y(t_0 + 2h), \ldots$$

by the recurrence

$$y_{n+1} = y_n + hf(t_n, y_n).$$

Systems of linear first order differential equations can be discretised exactly analytically using the methods shown in the discretisation article.

Applications

Biology: Some of the best-known difference equations have their origins in the attempt to model population dynamics.

For example, the Fibonacci numbers were once used as a model for the growth of a rabbit population.

The logistic map is used either directly to model population growth, or as a starting point for more detailed models. In this context, coupled difference equations are often used to model the interaction of two or more populations.

For example, the Nicholson-Bailey model for a host-parasite interaction is given by

$$N_{t+1} = \lambda N_t e^{-aP_t}$$

$$P_{t+1} = N_t(1 - e^{-aP_t}),$$

with N_t representing the hosts, and P_t the parasites, at time t.

Integrodifference equations are a form of recurrence relation important to spatial ecology. These and other difference equations are particularly suited to modelling univoltine populations.

Digital Signal Processing: In digital signal processing, recurrence relations can model feedback in a system, where outputs at one time become inputs for future time. They thus arise in infinite impulse response (IIR) digital filters.

For example, the equation for a "feedforward" IIR comb filter of delay T is:

$$y_t = (1-\alpha)x_t + \alpha y_{t-T}$$

Where x_t is the input at time t, y_t is the output at time t, and α controls how much of the delayed signal is fed back into the output. From this we can see that

$$y_t = (1-\alpha)x_t + \alpha((1-\alpha)x_{t-T} + \alpha y_{t-2T})$$

$$y_t = (1-\alpha)x_t + (\alpha-\alpha^2)x_{t-T} + \alpha^2 y_{t-2T}$$

etc.

Economics: Recurrence relations, especially linear recurrence relations, are used extensively in both theoretical and empirical economics. In particular, in macroeconomics one might develop a model of various broad sectors of the economy (the financial sector, the goods sector, the labour market, etc.) in which some agents' actions depend on lagged variables. The model would then be solved for current values of key variables (interest rate, real GDP, etc.) in terms of exogenous variables and lagged endogenous variables.

Chapter 13

Differential Equations

Singular Solutions

The theory of singular solutions of ordinary and partial differential equations was a subject of research from the time of Leibniz, but only since the middle of the nineteenth century did it receive special attention. A valuable but little-known work on the subject is that of Houtain (1854). Darboux (starting in 1873) was a leader in the theory, and in the geometric interpretation of these solutions he opened a field which was worked by various writers, notably Casorati and Cayley. To the latter is due (1872) the theory of singular solutions of differential equations of the first order as accepted circa 1900.

Reduction to Quadratures

The primitive attempt in dealing with differential equations had in view a reduction to quadratures. As it had been the hope of eighteenth-century algebraists to find a method for solving the general equation of the nth degree, so it was the hope of analysts to find a general method for integrating any differential equation. Gauss (1799) showed, however, that the differential equation meets its limitations very soon unless complex numbers are introduced. Hence analysts began to substitute the study of functions, thus opening a new and fertile field. Cauchy was the first to appreciate the importance of this view. Thereafter the real question was to be, not whether a solution is possible by means of known functions or their integrals, but whether a given differential equation suffices for the definition of a function of the independent variable or variables, and if so, what are the characteristic properties of this function.

Fuchsian Theory

Two memoirs by Fuchs (*Crelle*, 1866, 1868), inspired a novel approach, subsequently elaborated by Thomé and Frobenius. Collet was a prominent contributor beginning in 1869, although his method for integrating a non-linear system was communicated to Bertrand in 1868. Clebsch (1873) attacked the theory along lines parallel to those followed in his theory of Abelian integrals. As the latter can be classified according to the properties of the fundamental curve which remains unchanged under a rational transformation, so Clebsch proposed to classify the transcendent functions defined by the differential equations according to the invariant properties of the corresponding surfaces $f = 0$ under rational one-to-one transformations.

Lie's Theory

From 1870 Sophus Lie's work put the theory of differential equations on a more satisfactory foundation. He showed that the integration theories of the older mathematicians can, by the introduction of what are now called Lie groups, be referred to a common source; and that ordinary differential equations which admit the same infinitesimal transformations present comparable difficulties of integration. He also emphasised the subject of transformations of contact. Lie's group theory of differential equations, has been certified, namely: (1) that it unifies the many ad hoc methods known for solving differential equations, and (2) that it provides powerful new ways to find solutions. The theory has applications to both ordinary and partial differential equations.

A general approach to solve DE's uses the symmetry property of differential equations, the continuous infinitesimal transformations of solutions to solutions (Lie theory). Continuous group theory, Lie algebras and differential geometry are used to understand the structure of linear and nonlinear (partial) differential equations for generating integrable equations, to find its Lax pairs, recursion operators, Bäcklund transform and finally finding exact analytic solutions to the DE.

Symmetry methods have been recognised to study differential equations arising in mathematics, physics, engineering, and many other disciplines.

Sturm–Liouville Theory

Sturm–Liouville theory is a theory of eigenvalues and eigenfunctions of linear operators defined in terms of second-order homogeneous linear equations, and is useful in the analysis of certain partial differential equations.

Existence and Uniqueness of Solutions

There are several theorems that establish existence and uniqueness of solutions to initial value problems involving ODEs both locally and globally. The two main theorems are

Theorem	*Assumption*	*Conclusion*
Peano existence theorem	F continuous	local existence only
Picard–Lindelöf theorem	F Lipschitz continuous	local existence and uniqueness

which are both local results.

Local Existence and Uniqueness Theorem Simplified

The theorem can be stated simply as follows. For the equation and initial value problem:

$$y' = F(x, y), \quad y_0 = y(x_0)$$

if F and $-F/-y$ are continuous in a closed rectangle

$$R = [x_0 - a, x_0 + a] \times [y_0 - b, x_0 + b]$$

in the x-y plane, where a and b are real (symbolically: $a,\ b \in \mathbb{R}$) and × denotes the cartesian product, square brackets denote closed intervals, then there is an interval

$$I = [x_0 - h, x_0 + h] \subset [x_0 - a, x_0 + a]$$

for some $h \in \mathbb{R}$ where *the* solution to the above equation and initial value problem can be found. That is, there is a solution and its unique. Since there is no restriction on F to be linear, this applies to non-linear equations which take the form $F(x, y)$, and it can also be applied to systems of equations.

Global Uniqueness and Maximum Domain of Solution

When the hypotheses of the Picard–Lindelöf theorem are satisfied, then local existence and uniqueness can be extended to a global result. More precisely:

For each initial condition (x_0, y_0) there exists a unique maximum (possibly infinite) open interval

$$I_{max} = (x_-, x_+), x_\pm \in \mathbb{R}, x_0 \in I_{max}$$

such that any solution which satisfies this initial condition is a restriction of the solution which satisfies this initial condition with domain $I_{\max}$.

In the case that $x_\pm \nrightarrow \pm\infty$, there are exactly two possibilities

- explosion in finite time: $\lim_{x \to x_\pm} \|y(x)\| \to \infty$
- leaves domain of definition: $\lim_{x \to x_\pm} \in \partial\bar{\Omega}$

where Ω is the open set in which F is defined, and $\partial\bar{\Omega}$ is its boundary.

Note that the maximum domain of the solution

- is always an interval (to have uniqueness)
- may be smaller than $\mathbb{R}$
- may depend on the specific choice of (x_0, y_0).

Example:

$$y' = y^2$$

This means that $F(x, y) = y^2$, which is C^1 and therefore Lipschitz continuous for all y, satisfying the Picard–Lindelöf theorem.

Even in such a simple setting, the maximum domain of solution cannot be all $\mathbb{R}$, since the solution is

$$y(x) = \frac{y_0}{(x_0 - x)y_0 + 1}$$

which has maximum domain:

$$\begin{cases} \mathbb{R} & y_0 = 0 \\ (-\infty, x_0 + \dfrac{1}{y_0}) & y_0 > 0 \\ (x_0 + \dfrac{1}{y_0}, +\infty) & y_0 < 0 \end{cases}$$

This shows clearly that the maximum interval may depend on the initial conditions.

The domain of y could be taken as being $\mathbb{R} \setminus (x_0 + 1/y_0)$, but this would lead to a domain that is not an interval, so that the side opposite to the initial condition would be disconnected from the initial condition, and therefore not uniquely determined by it.

The maximum domain is not $\mathbb{R}$ because

$$\lim_{x \to x_\pm} \|y(x)\| \to \infty,$$

which is one of the two possible cases according to the above theorem.

Reduction of Order

Differential equations can usually be solved more easily if the order of the equation can be reduced.

Reduction to a First Order System

Any differential equation of order n,

$$F\left(x, y, y', y'', \cdots, y^{(n-1)}\right) = y^{(n)}$$

can be written as a system of n first-order differential equations by defining a new family of unknown functions

$$y_i = y^{(i-1)}.$$

for i = 1, 2,... n.

The n-dimensional system of first-order coupled differential equations is then

$$\begin{array}{ccc} y_{1'} & = & y_2 \\ y_{2'} & = & y_3 \\ & \vdots & \\ y_{n-1'} & = & y_n \\ y_{n'} & = & F(x, y_1, \cdots, y_n). \end{array}$$

more compactly in vector notation:

$$\mathbf{y}' = \mathbf{F}(x, \mathbf{y})$$

where

$$\mathbf{y} = (y_1, \cdots, y_n), \quad \mathbf{F}(x, y_1, \cdots, y_n) = (y_2, \cdots, y_n, F(x, y_1, \cdots, y_n)).$$

Software for ODE Solving

- FuncDesigner (free license: BSD, uses Automatic differentiation, also can be used online via Sage-server)
- odeint - A C++ library for solving ordinary differential equations numerically
- VisSim - a visual language for differential equation solving
- Mathematical Assistant on Web online solving first order (linear and with separated variables) and second order linear differential equations (with constant coefficients), including intermediate steps in the solution.
- DotNumerics: Ordinary Differential Equations for C# and VB.NET Initial-value problem for nonstiff and stiff ordinary differential equations (explicit Runge-Kutta, implicit Runge-Kutta, Gear's BDF and Adams-Moulton).
- Online experiments with JSXGraph
- Maxima computer algebra system (GPL)
- COPASI a free (Artistic License 2.0) software package for the integration and analysis of ODEs.
- MATLAB a matrix-programming software (MATrix LABoratory)
- GNU Octave a high-level language, primarily intended for numerical computations.

Partial Differential Equation

In mathematics, a partial differential equation (PDE) is a differential equation that contains unknown multivariable functions and their partial derivatives. PDEs are used to formulate problems involving functions of several variables, and are either solved by hand, or used to create a relevant computer model.

PDEs can be used to describe a wide variety of phenomena such as sound, heat, electrostatics, electrodynamics, fluid flow, or elasticity. These seemingly distinct physical phenomena can be formalised identically in terms of PDEs, which shows that they are governed by the same underlying dynamic. Just as ordinary differential equations often model one-dimensional dynamical systems, partial differential equations often model multidimensional systems. PDEs find their generalisation in stochastic partial differential equations.

Partial Differential Equations (PDEs) are equations that involve rates of change with respect to continuous variables. The configuration of a rigid body is specified by six numbers, but the configuration of a fluid is given by the continuous distribution of the temperature, pressure, and so forth.

The dynamics for the rigid body take place in a finite-dimensional configuration space; the dynamics for the fluid occur in an infinite-dimensional conûguration space. This distinction usually makes PDEs much harder to solve than Ordinary Differential Equations (ODEs), but here again there will be simple solutions for linear problems. Classic domains where PDEs are used include acoustics, fluid flow, electrodynamics, and heat transfer.

A partial differential equation (PDE) for the function $u(x_1,...x_n)$ is an equation of the form

$$F(x_1,\ldots,x_n,u,\frac{\partial u}{\partial x_1},\ldots,\frac{\partial u}{\partial x_n},\frac{\partial^2 u}{\partial x_1 \partial x_1},\ldots,\frac{\partial^2 u}{\partial x_1 \partial x_n},\ldots)=0.$$

If F is a linear function of u and its derivatives, then the PDE is called linear. Common examples of linear PDEs include the heat equation, the wave equation and Laplace's equation.

A relatively simple PDE is

$$\frac{\partial u}{\partial x}(x,y)=0.$$

This relation implies that the function $u(x, y)$ is independent of x. However, the equation gives no information on the function's dependence on the variable y. Hence the general solution of this equation is

$$u(x,y)=f(y),$$

where f is an arbitrary function of y. The analogous ordinary differential equation is

$$\frac{\mathrm{d}u}{\mathrm{d}x}(x)=0,$$

which has the solution

$$u(x)=c,$$

where c is any constant value. These two examples illustrate that general solutions of ordinary differential equations (ODEs) involve arbitrary constants, but solutions of PDEs involve arbitrary functions. A solution of a PDE is generally not unique; additional conditions must generally be specified on the boundary of the region where the solution is defined. For instance, in the simple example above, the function $f(y)$ can be determined if u is specified on the line $x = 0$.

Existence and Uniqueness

Although the issue of existence and uniqueness of solutions of ordinary differential equations has a very satisfactory answer with the Picard–Lindelöf theorem, that is far from the case for partial differential equations. The Cauchy–Kowalevski theorem states that the Cauchy problem for any partial differential equation whose coefficients are analytic in the unknown function and its derivatives, has a locally unique analytic solution.

Although this result might appear to settle the existence and uniqueness of solutions, there are examples of linear partial differential equations whose coefficients have derivatives of all orders (which are nevertheless not analytic) but which have no solutions at all. Even if the solution of a partial differential equation exists and is unique, it may nevertheless have undesirable properties. The mathematical study of these questions is usually in the more powerful context of weak solutions.

An example of pathological behaviour is the sequence of Cauchy problems (depending upon n) for the Laplace equation

$$\frac{\partial^2 u}{\partial x^2} + \frac{\partial^2 u}{\partial y^2} = 0,$$

with boundary conditions

$$u(x,0) = 0,$$

$$\frac{\partial u}{\partial y}(x,0) = \frac{\sin nx}{n},$$

where n is an integer. The derivative of u with respect to y approaches 0 uniformly in x as n increases, but the solution is

$$u(x,y) = \frac{(\sinh ny)(\sin nx)}{n^2}.$$

This solution approaches infinity if nx is not an integer multiple of π for any non-zero value of y. The Cauchy problem for the Laplace equation is called *ill-posed* or *not well posed*, since the solution does not depend continuously upon the data of the problem. Such ill-posed problems are not usually satisfactory for physical applications.

Notation

In PDEs, it is common to denote partial derivatives using subscripts. That is:

$$u_x = \frac{\partial u}{\partial x}$$

$$u_{xy} = \frac{\partial^2 u}{\partial y \partial x} = \frac{\partial}{\partial y}\left(\frac{\partial u}{\partial x}\right).$$

Especially in physics, del (∇) is often used for spatial derivatives, and $\dot{u}, \ddot{u}$ for time derivatives. For example, the wave equation (described below) can be written as

$$\ddot{u} = c^2 \nabla^2 u$$

or

$$\ddot{u} = c^2 \Delta u$$

where Δ is the Laplace operator.

Heat Equation in One Space Dimension

The equation for conduction of heat in one dimension for a homogeneous body has the form

$$u_t = \alpha u_{xx}$$

where $u(t,x)$ is temperature, and α is a positive constant that describes the rate of diffusion. The Cauchy problem for this equation consists in specifying $u(0, x) = f(x)$, where $f(x)$ is an arbitrary function.

General solutions of the heat equation can be found by the method of separation of variables.

Some examples appear in the heat equation article. They are examples of Fourier series for periodic f and Fourier transforms for non-periodic f. Using the Fourier transform, a general solution of the heat equation has the form

$$u(t,x)=\frac{1}{\sqrt{2\pi}}\int_{-\infty}^{\infty}F(\xi)e^{-\alpha\xi^2 t}e^{i\xi x}\,d\xi,$$

where *F* is an arbitrary function. To satisfy the initial condition, *F* is given by the Fourier transform of *f*, that is

$$F(\xi)=\frac{1}{\sqrt{2\pi}}\int_{-\infty}^{\infty}f(x)e^{-i\xi x}\,dx.$$

If *f* represents a very small but intense source of heat, then the preceding integral can be approximated by the delta distribution, multiplied by the strength of the source. For a source whose strength is normalised to 1, the result is

$$F(\xi)=\frac{1}{\sqrt{2\pi}},$$

and the resulting solution of the heat equation is

$$u(t,x)=\frac{1}{2\pi}\int_{-\infty}^{\infty}e^{-\alpha\xi^2 t}e^{i\xi x}\,d\xi.$$

This is a Gaussian integral. It may be evaluated to obtain

$$u(t,x)=\frac{1}{2\sqrt{\pi\alpha t}}\exp\left(-\frac{x^2}{4\alpha t}\right).$$

This result corresponds to the normal probability density for x with mean 0 and variance $2\alpha t$. The heat equation and similar diffusion equations are useful tools to study random phenomena. Wave equation in one spatial dimension.

The wave equation is an equation for an unknown function $u(t, x)$ of the form

$$u_{tt}=c^2u_{xx}.$$

Here u might describe the displacement of a stretched string from equilibrium, or the difference in air pressure in a tube, or the magnitude of an electromagnetic field in a tube, and c is a number that corresponds to the velocity of the wave.

The Cauchy problem for this equation consists in prescribing the initial displacement and velocity of a string or other medium:

$$u(0,x)=f(x),$$

$$u_t(0,x)=g(x),$$

where *f* and *g* are arbitrary given functions. The solution of this problem is given by d'Alembert's formula:

$$u(t,x) = \frac{1}{2}\left[f(x-ct)+f(x+ct)\right]+\frac{1}{2c}\int_{x-ct}^{x+ct} g(y)dy.$$

This formula implies that the solution at (t,x) depends only upon the data on the segment of the initial line that is cut out by the characteristic curves

$$x - ct = \text{constant}, \quad x + ct = \text{constant},$$

that are drawn backwards from that point. These curves correspond to signals that propagate with velocity c forward and backward. Conversely, the influence of the data at any given point on the initial line propagates with the finite velocity c: there is no effect outside a triangle through that point whose sides are characteristic curves. This behaviour is very different from the solution for the heat equation, where the effect of a point source appears (with small amplitude) instantaneously at every point in space.

The solution given above is also valid if t is negative, and the explicit formula shows that the solution depends smoothly upon the data: both the forward and backward Cauchy problems for the wave equation are well-posed.

Spherical Waves

Spherical waves are waves whose amplitude depends only upon the radial distance r from a central point source. For such waves, the three-dimensional wave equation takes the form

$$u_{tt} = c^2\left[u_{rr} + \frac{2}{r}u_r\right].$$

This is equivalent to

$$(ru)_{tt} = c^2\left[(ru)_{rr}\right],$$

and hence the quantity ru satisfies the one-dimensional wave equation. Therefore a general solution for spherical waves has the form

$$u(t,r) = \frac{1}{r}\left[F(r-ct)+G(r+ct)\right],$$

where F and G are completely arbitrary functions. Radiation from an antenna corresponds to the case where G is identically zero. Thus the wave form transmitted from an antenna has no distortion in time: the only distorting factor is $1/r$. This feature of undistorted propagation of waves is not present if there are two spatial dimensions.

Laplace Equation in Two Dimensions

The Laplace equation for an unknown function of two variables φ has the form

$$\varphi_{xx} + \varphi_{yy} = 0.$$

Solutions of Laplace's equation are called harmonic functions.

Connection with Holomorphic Functions: Solutions of the Laplace equation in two dimensions are intimately connected with analytic functions of a complex variable (a.k.a. holomorphic functions): the real and imaginary parts of any analytic function are conjugate harmonic functions: they both satisfy the Laplace equation, and their gradients are orthogonal. If f = u + iv, then the Cauchy–Riemann equations state that

$$u_x = v_y, \quad v_x = -u_y,$$

and it follows that

$$u_{xx} + u_{yy} = 0, \quad v_{xx} + v_{yy} = 0.$$

Conversely, given any harmonic function in two dimensions, it is the real part of an analytic function, at least locally. Details are given in Laplace equation.

A Typical Boundary Value Problem: A typical problem for Laplace's equation is to find a solution that satisfies arbitrary values on the boundary of a domain.

For example, we may seek a harmonic function that takes on the values $u(\theta)$ on a circle of radius one. The solution was given by Poisson:

$$\varphi(r,\theta) = \frac{1}{2\pi}\int_0^{2\pi} \frac{1-r^2}{1+r^2-2r\cos(\theta-\theta')} u(\theta')d\theta'.$$

Petrovsky (1967, p. 248) shows how this formula can be obtained by summing a Fourier series for φ. If $r<1$, the derivatives of φ may be computed by differentiating under the integral sign, and one can verify that φ is analytic, even if u is continuous but not necessarily differentiable.

This behaviour is typical for solutions of elliptic partial differential equations: the solutions may be much more smooth than the boundary data. This is in contrast to solutions of the wave equation, and more general hyperbolic partial differential equations, which typically have no more derivatives than the data.

Initial-boundary Value Problems

Many problems of mathematical physics are formulated as initial-boundary value problems.

Vibrating String: If the string is stretched between two points where x = 0 and x = L and u denotes the amplitude of the displacement of the string, then u satisfies the one-dimensional wave equation in the region where 0 < x < L and t is unlimited. Since the string is tied down at the ends, u must also satisfy the boundary conditions

$$u(t,0)=0, \quad u(t,L)=0,$$

as well as the initial conditions

$$u(0,x)=f(x), \quad u_t(0,x)=g(x).$$

The method of separation of variables for the wave equation

$$u_{tt}=c^2u_{xx},$$

leads to solutions of the form

$$u(t,x)=T(t)X(x),$$

where

$$T''+k^2c^2T=0, \quad X''+k^2X=0,$$

where the constant k must be determined. The boundary conditions then imply that X is a multiple of sin kx, and k must have the form

$$k=\frac{n\pi}{L},$$

where n is an integer. Each term in the sum corresponds to a mode of vibration of the string. The mode with n=1 is called the fundamental mode, and the frequencies of the other modes are all multiples of this frequency. They form the overtone series of the string, and they are the basis for musical acoustics. The initial conditions may then be satisfied by representing f

and g as infinite sums of these modes. Wind instruments typically correspond to vibrations of an air column with one end open and one end closed. The corresponding boundary conditions are

$$X(0) = 0, \quad X'(L) = 0.$$

The method of separation of variables can also be applied in this case, and it leads to a series of odd overtones. The general problem of this type is solved in Sturm–Liouville theory.

Vibrating Membrane: If a membrane is stretched over a curve C that forms the boundary of a domain D in the plane, its vibrations are governed by the wave equation

$$\frac{1}{c^2}u_{tt} = u_{xx} + u_{yy},$$

if $t>0$ and (x, y) is in D. The boundary condition is $u(t, x, y) = 0$ if (x, y) is on C. The method of separation of variables leads to the form

$$u(t,x,y) = T(t)v(x,y),$$

which in turn must satisfy

$$\frac{1}{c^2}T'' + k^2T = 0,$$

$$v_{xx} + v_{yy} + k^2v = 0.$$

The latter equation is called the Helmholtz Equation. The constant k must be determined to allow a non-trivial v to satisfy the boundary condition on C. Such values of k^2 are called the eigenvalues of the Laplacian in D, and the associated solutions are the eigenfunctions of the Laplacian in D. The Sturm–Liouville theory may be extended to this elliptic eigenvalue problem (Jost, 2002).

Other Examples

The Schrödinger equation is a PDE at the heart of non-relativistic quantum mechanics. In the WKB approximation it is the Hamilton–Jacobi equation. Except for the Dym equation and the Ginzburg–Landau equation, the above equations are linear in the sense that they can be written in the form $Au = f$ for a given linear operator A and a given function f. Other important non-linear equations include the Navier–Stokes equations describing the flow of fluids, and Einstein's field equations of general relativity.

Classification

Some linear, second-order partial differential equations can be classified as parabolic, hyperbolic or elliptic. Others such as the Euler–Tricomi equation have different types in different regions. The classification provides a guide to appropriate initial and boundary conditions, and to smoothness of the solutions.

Analytical Methods to Solve PDEs

Separation of Variables

Linear PDEs can be reduced to systems of ordinary differential equations by the important technique of separation of variables. The logic of this technique may be confusing upon first acquaintance, but it rests on the uniqueness of solutions to differential equations: as with ODEs, if you can find any solution that solves the equation and satisfies the boundary conditions, then it is the solution. We assume as an ansatz that the dependence of the solution on space and time can be written as a product of terms that each depend on a single coordinate, and then see if and how this can be made to solve the problem.

In the method of separation of variables, one reduces a PDE to a PDE in fewer variables, which is an ODE if in one variable – these are in turn easier to solve. This is possible for simple PDEs, which are called separable partial differential equations, and the domain is generally a rectangle (a product of intervals). Separable PDEs correspond to diagonal matrices – thinking of "the value for fixed x– as a coordinate, each coordinate can be understood separately. This generalises to the method of characteristics, and is also used in integral transforms.

Method of Characteristics

In special cases, one can find characteristic curves on which the equation reduces to an ODE – changing coordinates in the domain to straighten these curves allows separation of variables, and is called the method of characteristics. More generally, one may find characteristic surfaces.

Integral Transform

An integral transform may transform the PDE to a simpler one, in particular a separable PDE. This corresponds to diagonalising an operator. An important example of this is

Fourier analysis, which diagonalises the heat equation using the eigenbasis of sinusoidal waves.

If the domain is finite or periodic, an infinite sum of solutions such as a Fourier series is appropriate, but an integral of solutions such as a Fourier integral is generally required for infinite domains. The solution for a point source for the heat equation given above is an example for use of a Fourier integral.

Fundamental Solution

Inhomogeneous equations can often be solved (for constant coefficient PDEs, always be solved) by finding the fundamental solution (the solution for a point source), then taking the convolution with the boundary conditions to get the solution. This is analogous in signal processing to understanding a filter by its impulse response.

Superposition Principle

Because any superposition of solutions of a linear, homogeneous PDE is again a solution, the particular solutions may then be combined to obtain more general solutions.

Methods for Non-linear Equations

There are no generally applicable methods to solve non-linear PDEs. Still, existence and uniqueness results (such as the Cauchy–Kowalevski theorem) are often possible, as are proofs of important qualitative and quantitative properties of solutions (getting these results is a major part of analysis). Computational solution to the nonlinear PDEs, the split-step method, exist for specific equations like nonlinear Schrödinger equation.

Nevertheless, some techniques can be used for several types of equations. The h-principle is the most powerful method to solve underdetermined equations. The Riquier–Janet theory is an effective method for obtaining information about many analytic overdetermined systems.The method of characteristics (similarity transformation method) can be used in some very special cases to solve partial differential equations.

In some cases, a PDE can be solved via perturbation analysis in which the solution is considered to be a correction to an equation with a known solution. Alternatives are numerical analysis techniques from simple finite difference schemes to the more mature multigrid and finite element methods. Many interesting problems in science and engineering are solved in this way using computers, sometimes high performance supercomputers.

Lie Group Method

From 1870 Sophus Lie's work put the theory of differential equations on a more satisfactory foundation. He showed that the integration theories of the older mathematicians can, by the introduction of what are now called Lie groups, be referred to a common source; and that ordinary differential equations which admit the same infinitesimal transformations present comparable difficulties of integration. He also emphasised the subject of transformations of contact.

A general approach to solve PDE's uses the symmetry property of differential equations, the continuous infinitesimal transformations of solutions to solutions (Lie theory). Continuous group theory, Lie algebras and differential geometry are used to understand the structure of linear and nonlinear partial differential equations for generating integrable equations, to find its Lax pairs, recursion operators, Bäcklund transform and finally finding exact analytic solutions to the PDE. Symmetry methods have been recognised to study differential equations arising in mathematics, physics, engineering, and many other disciplines.

Semianalytical Method

The adomian decomposition method, the Lyapunov artificial small parameter method and He's homotopy perturbation method are all special cases of the more general homotopy analysis method. There are series expansion method but there are independent of small physical parameters compared to thee well known perturbation theory.

Numerical Methods to Solve PDEs

The three most widely used numerical methods to solve PDEs are the finite element method (FEM), finite volume methods (FVM) and finite difference methods (FDM). The FEM has a prominent position among these methods and especially its exceptionally efficient higher-order version hp-FEM. Other versions of FEM include the generalised finite element method (GFEM), extended finite element method (XFEM), spectral finite element method (SFEM), meshfree finite element method, discontinuous Galerkin finite element method (DGFEM), etc.

Finite Element Method

The finite element method (FEM) (its practical application often known as finite element analysis (FEA)) is a numerical technique for finding approximate solutions of partial differential equations (PDE) as well as of integral equations. The solution approach is based either on eliminating the differential equation completely (steady state problems), or rendering the PDE into an approximating system of ordinary differential equations, which are then numerically integrated using standard techniques such as Euler's method, Runge–Kutta, etc.

Finite Difference Method

Finite-difference methods are numerical methods for approximating the solutions to differential equations using finite difference equations to approximate derivatives.

Finite Volume Method

Similar to the finite difference method or finite element method, values are calculated at discrete places on a meshed geometry. "Finite volume" refers to the small volume surrounding each node point on a mesh. In the finite volume method, volume integrals in a partial differential equation that contain a divergence term are converted to surface integrals, using the divergence theorem. These terms are then evaluated as fluxes at the surfaces of each finite volume. Because the flux entering a given volume is identical to that leaving the adjacent volume, these methods are conservative.

Bibliography

Adomian, George, *Nonlinear Stochastic Operator Equations,* Academic Press Inc., Orlando, FL, 2006.

________, *Stochastic Systems: Mathematics in Science and Engineering,* Academic Press Inc., Orlando, FL, 2003.

Ang, W. T.: *A Beginner's Course in Boundary Element Methods*, Universal Publishers, 2023.

Banerjee, P. K.: *The Boundary Element Methods in Engineering*, McGraw-Hill College, 2014.

Birkhoff, G and Rota, G., *Ordinary Differential Equations*, Blaisdell, New York, 2002.

Butcher, J., *Numerical Methods for Ordinary Differential Equations*, Wiley, Hoboken, NJ, 2020.

Bernt K., *Stochastic Differential Equations: An Introduction with Applications*, Springer, Berlin, 2018.

Fitts, C. R., *Groundwater Science*, Elsevier/Academic Press, San Diego, CA, 2012.

Fornberg, B., *Practical Guide to Pseudospectral Methods*, Cambridge University Press, New York, NY, 2016.

Gardiner, C. W.: *Handbook of Stochastic Methods: for Physics*, Springer, 2014.

Gear, C. W., *Numerical Initial Value Problems in Ordinary Differential Equations,* Prentice Hall, Englewood Cliffs, NJ, 2001.

Gernot Beer, Ian Smith, Christian Duenser, *The Boundary Element Method with Programming: For Engineers and Scientists*, Springer, 2018.

Gibson, Walton C., *The Method of Moments in Electromagnetics*, Chapman & Hall/CRC, 2018.

Haitjema, H. M., *Analytic Element Modeling of Ground Water flow,* Academic Press, San Diego, CA., 2015.

Henrici, P., *Discrete Variable Methods in Ordinary Differential Equations*, John Wiley & Sons, New York, NY, 2002.

Hirsch, M. W., and Smale, S., *Differential Equations, Dynamical Systems and Linear Algebra*, Academic Press, New York, 2000.

Isaacson, E., and Keller, H. B., *Analysis of Numerical Methods*, Dover, New York, 2014.

Iserles, A., *A First Course in the Numerical Analysis of Differential Equations*, Cambridge University Press, Cambridge, 2016.

John T. Katsikadelis, *Boundary Elements Theory and Applications*, Elsevier, Oxford, 2023.

Magid, Andy R., *Lectures on Differential Galois Theory*, American Mathematical Society, Providence, R.I., 2014.

Platen, E.: *Numerical Solution of Stochastic Differential Equations*, Springer, 2015.

Strack, O. D. L., *Groundwater Mechanics*, Prentice Hall, Englewood Cliffs, NJ., 2009.

Teugels, J. and Sund B., *Encyclopedia of Actuarial Science*, Wiley, Chichester, 2014.

Thambynayagam, R. K. M.: *The Diffusion Handbook: Applied Solutions for Engineers*, McGraw-Hill, New York, 2019.

Thomas Mikosch, *Elementary Stochastic Calculus: with Finance in View*, World Scientific Publishing, Singapore, 2008.

Van Der Put, Marius; Singer, Michael F., *Galois Theory of Linear Differential Equations*, Springer-Verlag, Berlin, New York, 2013.

Wrobel, L. C.; M. H. Aliabadi: *The Boundary Element Method*, New Jersey, Wiley, 2016.

Index

H

I

L

M

N

P

Q

R

S

T

W

❑❑❑